Architecture in the Making

Rita Salamouni · Nicola Scardigno ·
Giuseppe Strappa

Architecture in the Making

Conversations on Urban Morphology and Design

Springer

Rita Salamouni
Department of Architecture
and Construction
La Sapienza University
Rome, Italy

Nicola Scardigno
Department of Architecture
Politecnico di Bari
Bari, Italy

Giuseppe Strappa
Department of Architecture
Roma Tre University
Rome, Italy

ISBN 978-3-031-86832-0 ISBN 978-3-031-86833-7 (eBook)
https://doi.org/10.1007/978-3-031-86833-7

Revised and expanded translation from the Italian language edition: "Forma in divenire" by Nicola Scardigno, © Giuseppe Strappa 2023. Published by Letteraventidue. All Rights Reserved.

This Springer imprint is published by the registered company Springer Nature Switzerland AG
The registered company address is: Gewerbestrasse 11, 6330 Cham, Switzerland

Foreword by Franco Purini

A Treatise in the Form of a Dialogue

When compared to the conversations one can see and hear on television programs, the written interviews found in newspapers, magazines, and books remain in the mind in a lasting way. In fact, for many years now, the interview has become a significant, complex, and widespread literary form. What is asked and the answers to the questions imprint themselves on the reading, forming a permanent message.

There are various ways of interviewing. One can choose the theme of biography, which recounts a life in its varied unfolding. One can ask about one's opinions on accidental events, mistakes, or positive things one has done. Sometimes it is mainly asked about the context in which one lived and the influence it had on us. In other conversations, the main topic is the relationship with knowledge other than the one we have chosen and cultivated. A description of our qualities is also frequent in interviews. Finally, they can have, as the focus of the discussion, the expression of a theory. These are, in my opinion, the most culturally incisive ideas that presuppose communication skills as well as the ability to expound one's beliefs.

It must also be said that the difference between interview and dialogue has long since disappeared. The "Interviste impossibili" (impossible interviews) of the 1970s, together with the "Alle otto della sera" (At eight o'clock in the evening) conversations in the following period, have made the talk of two people a much more cultured discursive sphere, in which dialogue—a term that indicates a more authentic and profound encounter than a simple exchange of news and opinions—often deals with problems, orders and reasoning that go beyond entertainment to take the form of real lectures. Obviously, the different forms of the interview depend, to be understood as a whole, on the questions asked. Those who formulate them must not only be clear and precise but also capable, like a director, of conducting the interview with great rigor. Nicola Scardigno's and Rita Salamouni's work in their dialogues with Giuseppe Strappa is, from the point of view of content and the rhythm of the conversation, exemplary. What is asked in the questions proposed is a progressive narrative in which the individuality of who replies gives way to a system of linked notions defining the consequential level of choices. The set of answers to the questions, posed with excellent logic by Giuseppe Strappa's interlocutor, outlines a vast, precise and inspired conceptual framework. In short, this book is nothing other than a treatise on architecture proposed and articulated with wisdom and a mathematical progression, the result of the long and active teaching of the Roman teacher in the capital, in Bari and in many other cities, where he brought his fruitful knowledge. His treatise recalls historical Platonic dialogues, evokes Paul Valéry's "Eupalino", recalls the precious syntheses of Le Corbusier and the compositional magic of Mies van der Rohe, but above all follows the theoretical path that goes from Saverio Muratori to Gianfranco Caniggia. All this to define a system of statements that do not so much confirm previous references, as list a series of current, operationally precise, ideas produced by an assiduous study of the changes that morphology and typology have undergone in recent decades.

Giuseppe Strappa, therefore, does not validate the notions of typology and morphology present in the conceptions of Saverio Muratori and Gianfranco Caniggia, but identifies a new path toward a necessary innovation of these founding categories. An innovation that considers how much the evolution of the city in recent decades has changed the very meaning of urban settlements, in which informality, or the free arrangement of parts of the city, has taken the place of their previous physical-symbolic structure. This does not mean, moreover, that the characteristics of the suburbs of recent decades have no meaning, but these are no longer unitary, configuring themselves as the pieces of a vast mosaic that should be analyzed and interpreted with new paradigms.

It should also be noted that this treaty is not limited to being just a coherent system of general scientific rules. It is, in fact, pervaded by a humanistic atmosphere for which architecture is not the outcome of its genesis and its life over time, but a feeling imbued with the creative knowledge of a community. Reading the city like a book, even a book of poetry, is undoubtedly a precedent for Giuseppe Strappa's research in which the existence of the multitude is the authentic essence of living. His treatise, supported by a utopian vision, is something open, that is not immobile and constantly itself, but capable of foreseeing the incessant mutations of constructing. A treaty that is constantly evolving but, at the same time, always the same.

Nestled between these introductory texts and the concluding text by Matteo Ieva, preceded by some drawings by the protagonist of this agile volume, the book brings architecture back to its origins, removing it from any misunderstanding.

To conclude, the title of the book alludes to architecture understood as a world of elements to be composed into social groups, so to speak, not merely mechanical, or simply ordered into systems or sets. Living elements, we must remember, like the architecture that they will form, constantly mobile, waiting to make architecture what will be, after a certain time, its own memory and the prophecy of a future. Only this evolutionary nature can confirm in every season, in a positive contradiction, the primary and lasting meaning of living.

September 2024

Franco Purini
Sapienza University of Rome
Rome, Italy

Foreword by Jörg Gleiter

Architecture as Medium and Expression of Human Freedom

This book groups together both an interview with Giuseppe Strappa and his projects. It combines practice and theory, teaching and research. Overlapping like a palimpsest, they form a dense compilation of thoughts and intuitions. Using the expression of the author, this work is a substratum resulting from a long reflection on architecture, practice, story, and theory. The book has the potential to fully enter the debate on architectural design tools and have a long-term impact.

Strappa always has the whole picture in mind. In fact, architecture has an effect on life, just as life has an effect on architecture. It would be a misunderstanding to reduce it to just a material object. In other words, architecture is always the process of becoming.

In Strappa's definition, the *substratum* is what denotes the human condition, which has not always existed, but has accumulated over many centuries of culture, emerging through processes of metamorphosis, amalgamating with other parallel processes and finally into the current condition.

This substratum, however, is not simply lying in front of our eyes; on the contrary, each generation should appreciate it according to the spirit of its time, its own experience and with its own understanding of things.

Transmitting awareness and knowledge of this condition is the task of university teaching. It will then be up to the designer to assume responsibility.

The author's architectural projects bear witness to this. What Strappa wants to tell us in this book is that the architectural knowledge cannot simply be recalled: it must be continually re-appropriated, translated into new terms according to its historical conditions, making it fruitful for the present. One could also say that the knowledge of architecture must always be "infected" by the germs of the present time. Only in this way, can it develop its potential for the future.

The pre-Socratic philosopher Anaxagoras stated that man is an intelligent animal because he has hands and knows how to use them. This makes him a free being. In other words, the human freedom arises from the active action or by using the expression of the Roman architect, from the transformation of the matter into the material, this deriving from the morphological processes of formation. In this process, the form is the result of a critical reading and interpretation in which "reading and design" intertwine.

The projects presented in this book show Strappa's compassion for things and life. Architecture is based on this sympathy, as he vigorously shows us. Hence, Strappa's strong and fixed conviction that architecture is the means and evolving expression of human freedom. His work fits, I believe, into forms that evolved under different influences in a free society.

October 2025

Jörg Gleiter
Technical University of Berlin
Berlin, Germany

Acknowledgements

This work was supported by the Kaebup Erasmus+ European Union program [Grant Agreement No: 621585-EPP-1CY-EPPKA2-KA]. The European Commission's support for this publication does not constitute an endorsement of the contents, which only reflect the views of the authors.

The case studies explaining the work and methodology of Prof. Giuseppe Strappa are edited by Rita Salamouni.

Notes on Published Projects

Competition for the arrangement of Piazza Cairoli in Rome, 1985
First prize project
G. Strappa (group leader), G. Ceccarelli, R. Ciarnò

Proposal for the completion of Ca' Vernier dei Leoni in Venice. 1985
Selected project
Project for the 3rd international Architecture Exhibition at the Venice Biennale

New extension of the Terni Cemetry, second phase, 2005-2022
First prize project in the 1987 national competition
G. Strappa (group leader), T. Casatelli, P. Di Giuliomaria, E. Timpani

Proposal for a museum dedicated to Giorgio de Chirico, 1988
Drawings for an architectural exhibition at the invitation of the Academia di San Luca in Rome
G. Strappa (group leader), T. Casatelli, P. Di Giuliomaria, D. Guzzi

International competition "Vivienda y ciudad" for the landscaping of the southern part of Barcelona's Diagonal, 1989
G. Strappa (group leader), P. Di Giuliomaria, coll.: A. Trombetta

Competition for the construction of the new Laurentino Cemetry in Rome, 1990
First prize project
G. Strappa (group leader), P. Di Giuliomaria

Project for the exhibition "Silice e fuoco" at the Palazzo delle Esposizioni in Rome, 1992
G. Strappa (group leader), G. Mercurio, A. Valenti, M. Valenti

Project for a Church and parish services in the Tor Tre Teste area of Rome, 1995
G. Strappa (group leader), G. L. Maffei, T. Casatelli, P. Di Giuliomaria, A. Trombetta

Competition for the renovation of Piazza Mastai in the Trastevere, Rome, 1996
First prize project
G. Strappa (group leader), A. Trombetta and F. Ricordy (collaborator)

Competition for the redevelopment of Piazza San Cosimato in the Trastevere district of Rome, 2002
G. Strappa (group leader), M.L. Tavella, G. Carbonara (consultant), R. Bellardini (computist), E. Fortunati (collaborator)

Competition for the new municipal headquarters of Santa Marinella (Rome), 2004
G. Strappa (group leader), T. Casatelli, A. Franchetti Pardo, F. Maraglioli, A. Schietroma, M.L. Tavella

Competition for the expansion of the Home for the Elderly in Trani, 2005
National competition
G. Strappa (group leader), M. Ieva, C. Robbe

Updating of the Zen Quarter in the Palermo hinterland, 2008
Project at the invitation by the Faculty of Architecture of Palermo
G. Strappa (project leader), Alfonso Giancotti, Luca Reale, Federico De Matteis and Francesco Cianfarani (tutors), S. Di Benedetto, N. Dileo, M. Filosa,M. Pattarini, G. Papaevangeliu, L. Porqueddu, E. Prete (PhD students)

Competition for the redevelopment of an area in the historic center of Carezzano (Alessandria), 2011
First prize project
G. Strappa (group leader), A. Camiz, P. Carlotti, G. Galassi, M. Longo, M. Maretto

Design of new infrastructure and urban fabric in Partinico (Palermo), 2014
Proposal at the invitation by the Faculty of Architecture of Palermo
G. Strappa (group leader), P. Carlotti, A. Giancotti, L. Reale and A. R. Amato (tutors), C.Bernardini, F. Fava, A. Jelic (PhD students)

Reuse of the Expo 2015 area in Milan, 2015
Proposal at the invitation by the Politenico di Milano

Draco, Sapienza University of Rome
G. Strappa (coordinator); spatial study: P. Carlotti, I. Taci, C. Tartaglia; study of green areas: D. Nencini, V. Buongiorno; study of urban knots: G. Ciotoli, M. Falsetti; study of vertical fabric: I. Del Monaco, V. Mattei; basement study: P. Posocco, M. Raitano, G. Valeri

Project for the urban knotting of the Quebec Parkiament, 2017
Proposal developed within research carried out with the University of Laval (Quebec City)
G. Strappa (group leader), P. Carlotti, V. Buongiono, G. Emmi, C. Sammarco

Competition for an urban park in the area of the former central train station in Mendoza, 2018
Selected project
G. Strappa (group leader), C. Sammarco, S. Aloisio

Competition for the urban renewal of the municipality of Fara Olivana con Sola, 2022
Selected project
G. Strappa (group leader), M. Ieva, F.D. Rosa, C. Robbe

Thesis Projects

Reading and design of the city of Valletta, Politecnico di Bari, 2004
Thesis by students: A. D'Amato, C. Lapolla, L. Maggio, N. Minafra, G. Nanna, M. D. Piccininni
Tutor: G. Strappa, Co-tutor: M. Ieva

Typology of the urban organism of Trani within the Lombard wall, Politecnico di Bari, 2006
Thesis by students: F. De Benedectis, O. di Chito, E.a Gabriele, N. Incampo, L. Miano, R. Petrlli.
Tutor: G. Strappa, Co-tutors: M. Ieva, F. Onesti

Redesign of the Legislative Assembly Building in Plaza Tomè de Souza, Salvador de Bahia, as an urban node, Politecnico di Bari, 2006
Thesis by students: Z. Iacopelli, L. Lionetti, P. loizzo, M. A. Maggiore, L. Marvulli, A. I. Membola.
Tutor: G. Strappa, Co-tutor: M. Ieva.

Reading and design of the area of the convent Sv. Mergèles Marijos in Vilnius, Sapienza, Rome University, 2009
Thesis by the student F. Piazzoni.
Tutor: G. Strappa, Co-tutor: A. Franchetti Pardo.

Contents

About the Authors

Rita Salamouni is a French/Lebanese conservation architect. After studying architecture at the Lebanese American University, she obtained her master degree in conservation in Rome. She is currently enrolled in the Ph.D. program of Architecture and Construction of the city at La Sapienza University of Rome. She is particularly interested in studying the forming process of traditional of houses along the Mediterranean and the Red Sea, using both historical and morphological methodologies. She has also collaborated and worked with several international firms between Saudi Arabia, Lebanon, Italy and France. Her work mainly consists in studying different old settlements in order to understand their forming process and proceed with an adequate methodology for their conservation and restoration. Her activity mainly focuses on studying UNESCO heritage and historical sites in Saudi Arabia. Her aim is to combine both the morphological approach and the conservation strategies, trying to apply these methodologies in her work.

Nicola Scardigno is an architect with a Ph.D. in architectural and urban design and RtDa at the ArCoD Department of the Polytechnic of Bari.

Here he teaches the courses of "Architectural Design 1" and "Building Typologies and Urban Morphologies." He is a member of ISUFitaly and of the editorial committee of the U+D urban form and design magazine. He is the winner of international scholarships awarded by organizations (Isuf International) and universities (University of Wales). He participates in international research activities (Kaebup_Knowledge alliance for evidence-based urban practice), design competitions, national and international conferences and seminars. He carries out research on urban morphology and on issues affecting landscape transformations. He focuses his design activity, both educational and professional, and his writings on this area of interest.

Giuseppe Strappa architect, designer, and professor working in Rome, is currently a "high qualification" lecturer in Urban Morphology at the Faculty of Architecture of Roma Tre. He has carried out teaching and research activities at the University Sapienza of Rome, the Polytechnic University of Bari (where he was director of the

Department of Civil Engineering and Architecture) and at Université Laval in Québec City. At the Sapienza University, he was a full professor, member of the University Commission for Doctorates, coordinator of the Draco Doctorate in Architecture and Construction, director of the Doctoral School in Architectural Sciences, and director of the Lpa (Reading and Design of Architecture) Laboratory.

He is a Council member of Isuf, an international association dedicated to urban morphology, and coordinator of the IRN, Network of national urban morphology associations. He is editor of U+D, Urbanform and Design, and member of the Editorial Board of Urban Morphology.

He is the author of numerous publications and articles in national and international specialized journals and has contributed to newspapers of national importance (Repubblica, Corriere della Sera). For the FrancoAngeli publishing house in Milan, he directs the "Lettura e progetto" series. He has given lectures and conferences in Italian and foreign universities. He also carried out design activities since 1972, winning numerous architectural competitions. The focus of his activity, as a professor and designer, is the study of urban form read through its formative processes, considering the architectural project as the last, synthetic phase of an unfolding in continuous becoming.

He has carried out many projects, realizing some of them, has won numerous design competitions, and has received national and international recognition for his design research activities.

Some news and texts about his activity can be found at the website: www.giuseppestrappa.it

Part I
Conversations with Giuseppe Strappa

Chapter 1
Introduction

1.1 Space or Art of Delimitation

Combining both internal and external spaces, architecture is, *par excellence*, the art of delimitation. In this sense, the work of the Roman architect Giuseppe Strappa allows us to make some reflections on ways of conceiving the relationship between *space* and *limit*. Two notions deeply rooted in history as the "true form" of reality and which, in the opinion of the author, contribute to the construction of a lexicon necessary to also express architecture poetically.

The creative environment of Strappa's formation is the Roman one in the seventies, animated by the crisis of modernity. They were the years of the irruption of the dazzling research of Louis Kahn and of the innovative experiments of GRAU (Roman Group of Architects and Urbanists), especially through the original design research of his friend Alessandro Anselmi, its founder.

From here begins an intellectual journey particularly marked by the meeting with Gianfranco Caniggia. It will tend to investigate the "logical antecedent" of the figurative conditions of space, trying to grasp the meaning and the formative potential of the architecture concrete elements: construction systems, tectonic nodes, materials. The main aim seems to be that of freeing the logical and evidence-based aspects of the project from perception based aesthetic concepts. Then, he tries to demonstrate how the need to recover the architecture artistic center lies in the search for a transmissible, open and multiple links between the structural morphologies and the spatial forms related to them.

In particular, the search for the expressive value of architecture seems to materialize, in Strappa's work, through the development of an idea of an internal space (virtually open, even if covered) as the result of the gestures of "enclosing" and "knotting." Therefore, a paradigm of space intended as a built reality endowed with its own vigorous corporeity, to the point of configuring itself as the supporting nucleus of the architecture. This variation of the concept of space refers to the way in which the power of the sound of music fills silence or, I would say, to a Roman way of

R. Salamouni et al., *Architecture in the Making*,
https://doi.org/10.1007/978-3-031-86833-7_1

enhancing the expressive character of a spatial "body." It becomes representative of a plastic architectural form basically unitary, massive, strongly material, manifested through the semantics of the *limes* (Latin for limit). In implicit form, it is probably the latter, the true protagonist of Strappa's architecture: an evident, delimited and complex place in continuous formation, able to explain the typological, static and expressive questions of the architectural organism.

In the architect's work there seems to appear three ways to decline describe the relationship between space and limit.

The project of expanding the Terni Cemetery is precisely identified with the sign of an inhabited perimeter located on the border between the countryside and the "*eternal garden*" (Fig. 1.1). It is a metaphysical place, also known as the latter, where the Latin culture has always preserved the memory of its ancestors. Also, it evokes the model of the ancient *hortus conclusus* (closed orchard). According to the technique of bas-relief and excavation, the fields are in the area where three craters conceived, as the tradition of the Etruscan building world. They were cultivated by work and burials, hence the attribution of the term "good soil." The craters that appear within are in turn placed to contain a wood almost as if to evoke both the collective embrace of the chorus of sorrow and the female form, in alluding to the mother earth destined to generate. This large open natural space, representative of the agricultural cycle as a metaphor for human life, is contained within a long and thick building located in the shelter of a large roof. A veritable boundary built to "contain" the stone blocks of the private chapels and burials, to "be walked through" along the portico and at the same time become a place of "sight." In short, an architectural form which in its development following a high tripartite structure, marked by the alternating use of concrete and stone, cast iron and steel, gives a powerful symbolic value to the large plastic spatial reservoir. At the same time, it delimits and welcomes inside (through the intercolumns of the pseudo colonnade placed in combining both internal and external spaces, architecture is, *par* the central band of the surrounding wall) in the form of a dimly lit space. In fact, it is a question of two spaces that interpenetrate and belong to each other: that of the garden reservoir, being a synthesis between anthropic and natural forms and whose spatial reason lies precisely in the peremptoriness of the wall-fence sign; and that of the dense, dark, infinite longitudinal spatial body that runs inside the inhabited perimeter. A space, the latter, which is configured as a vacant material in extension to the garden and which coincides with the "processional way" placed between the theory of vertical white cylinders, the family chapels and the trussed roof with a metal structure (as in the Terni tradition) detached from the underlying wall. Almost as if to simultaneously declare the linguistic autonomy of the covering element, the legibility of the dilatation of the underlying path, but above all, the analytical composition of the thick longitudinal body. Here, in fact, the tripartite diaphragm wall, configuring itself as an element interposed between two different spaces (the one inside the longitudinal architectural body and that of the external space of the sacred courtyard), defines at the same time the reasons without denying that datum of co-incidence that places them on the level of dialectical comparison.

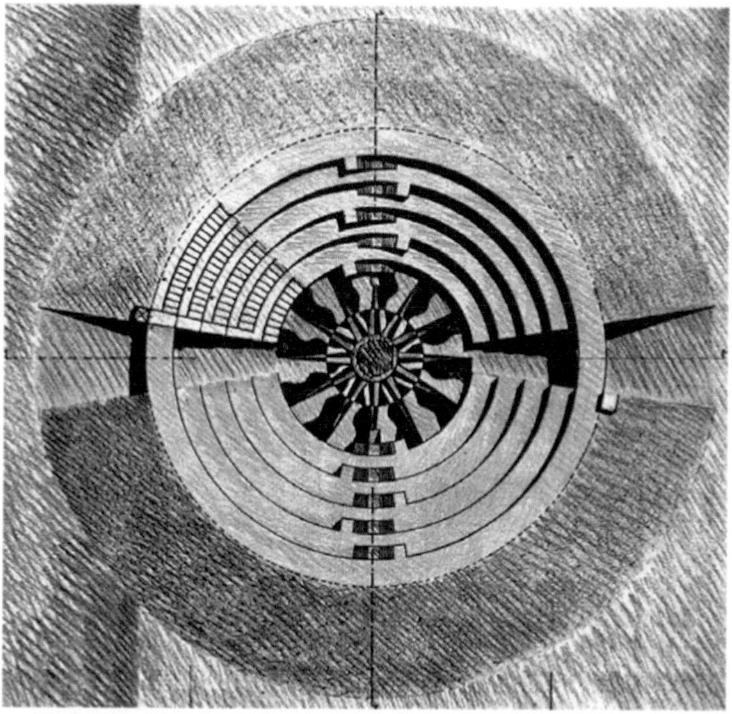

Fig. 1.1 Studies for a crater in Terni Cemetery, 1987

Within the same project, the relationship between limit and space is accomplished differently through a complete dematerialization of the internal space of the chapel where the Roman architect has envisaged the use of the concrete barrier wall.

The second way of describing the relationship between space and limit takes the form of the transition of an open space into a covered space. In the project for the current Québec Parliament Building, the design idea can be summarized using the Roman architect's original notion of the "knotting" (Fig. 1.2). An ideational process that underlies the formation of many building types characterized by the presence of a large central room served by serial rooms on the perimeter. In the case of the Canadian project, the large room is conceived as a new space available to the community, reached by urban routes that intersect within it, penetrating the body of the enclosure. In fact, it is design research clearly aimed at defining an internal, collective spatial quality through the elementary gesture of covering, starting from the recognition of a continuous and homogeneous liminal space. Of an already existing boundary: a *corpo triplo* (triple body) XIX-century building revolving on four sides organized with a central distribution space serving an enfilade of windowed rooms. In other words, a *limit* that corresponds to a continuous inhabited body, interposed between the city and the large courtyard inside the quadrangular construction. It is designed to solve the static problem of a unitary and solid space, which is to be concluded with a large dome in steel and polycarbonate. In short, a type of architecture that intends to highlight the relevance of an inner organic vaulted space precisely through the recognition of the expressive power of its own limits. It coincides with the core of the building. Explaining the relationship between the two elementary gestures of architecture, *fencing* and *covering*, which is enhanced in the search for congruent tectonics. It starts from the most genuine Roman architecture always capable of activating a formative dialectic between enclosure and spatial node by organically linking the roofing system to the perimeter structures that surround it. In the context of two of the four areas obtained from the intersection of urban routes, two spatial sub-conditions: one "positive," corresponding to a service block, the other "negative," corresponding to a small theater space circular plan. Two "objects" which by position, size and geometry declare their condition of punctual places containing spatial matrix. What emerges clearly is the theme of the large, covered space which, by gathering elements inside it, allows them to be free to appear, to establish relationships between each other and with the existing. As evidence of this compositional principle, there are designed two high-altitude walkway-routes having positions independent of the urban routes on the ground floor, to place the four existing bodies constituting the enclosure in relation to each other on the upper levels. Here, then, is how the large, covered space is in some way related to the places contained within it (the urban routes, the service block, the theater, the high-altitude connections) since, rather than being conceived as a void/absence, can be traced back to the great spatial node that takes shape inside its borders.

The third way of describing the relationship between space and limit takes the form of the project for the de Chirico Museum (Fig. 1.3). It is a visionary proposal designed in Rome for an exhibition at the Accademia di San Luca in 1988. This matrix of ideas that will be brought to synthesis in subsequent experiments, such as in the large corner tower in the Terni Cemetery itself. In other words, the idea around which the project revolves is that of a masonry building with a central plan and a powerful internal open space, marked in the middle by the staircase. In order to contain this space (conceived as declinations of the infinite through the exaltation of an inverted pyramid whose apex continues, virtually, to the center of the earth) is placed a *limes*—an enclosure with a thickness appropriate to the size of the place in which the enclosure is made of ramps and suspended galleries.

By taking place around the central space, this place tends to take the form of an enclosing spatial mass, of a "chiaroscuro," chromatic and iridescent nature in its upward development.

To facilitate the connotation of a totally internal spatial type within the dimension of the limit-envelope, the need of using a frame conceived as a geometric skeleton helps. A discrete system which, although interposed between the two main spatial bodies, openly declares its static belonging to the revolving body of the envelope and defines its conclusion.

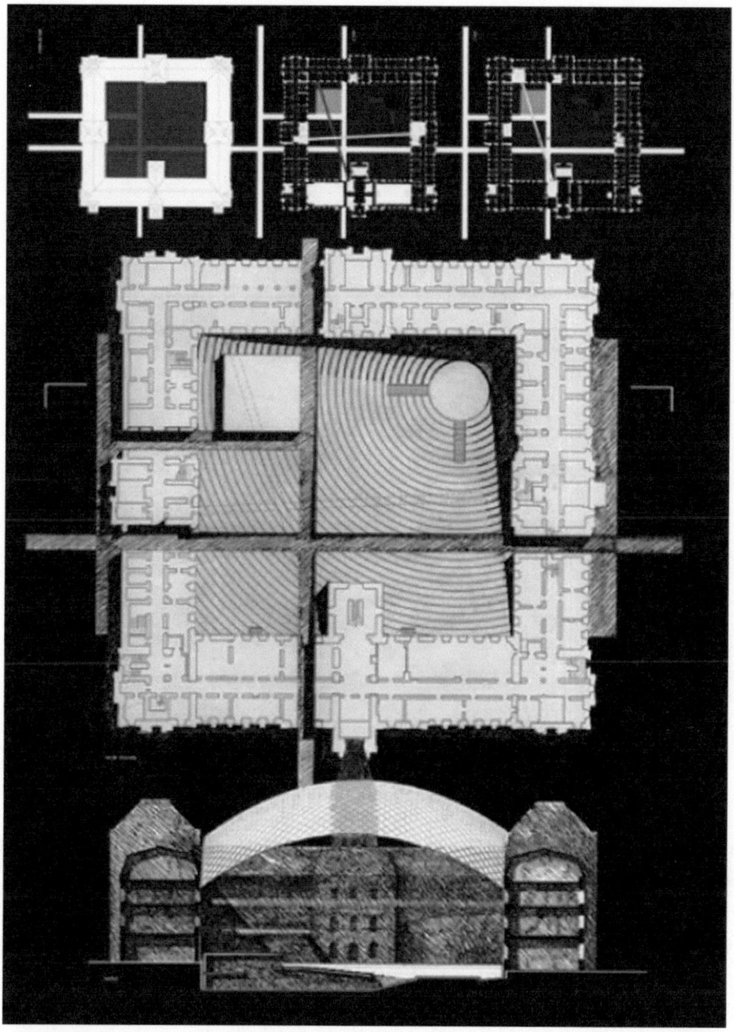

Fig. 1.2 Proposal for the transformation of the Quebec Parliament Building as an "urban knotting", 2017

1.2 Understanding the Territory

One of the fundamental assumptions in the idea of architecture proposed by Giuseppe Strappa is that its center is the form: this is what the designer must deal with, from which he derives the methods of both investigation and design.

This idea concerns all aspects of architectural design, starting with the most general and vast of all, the territorial one. What interests the architect about the territory is its form understood as the perceptible feature of a structure: through form

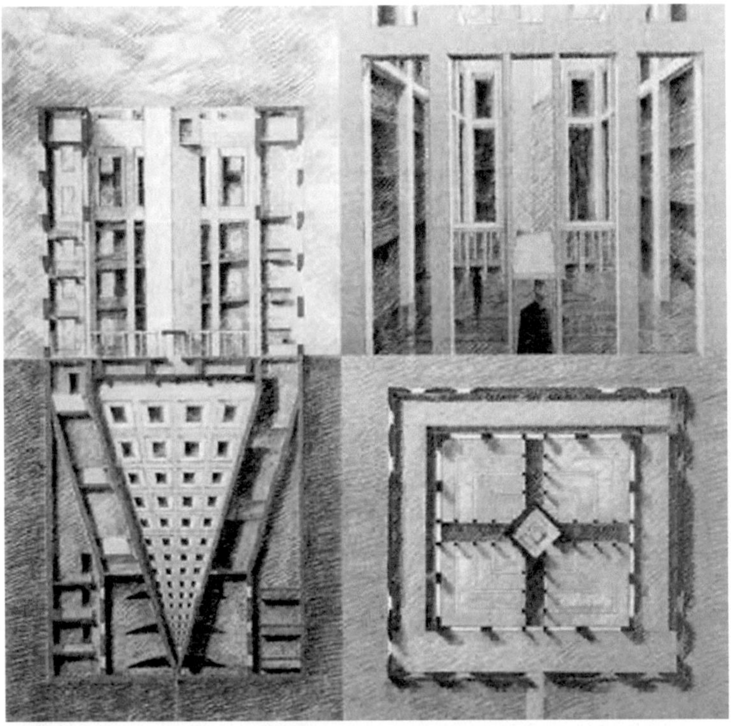

Fig. 1.3 Proposal for a museum dedicated to Giorgio de Chirico, 1988

the architect therefore reads the territory, its characters, and the processes that forms and designs it.

Territory is consequently a man-made construction that transforms nature, in the same way that a building can be considered a finalized transformation of matter.

This being one of his most constant convictions in design, Strappa explicitly shows in many of his project's reflections linked to the territorial interpretation of the environmental context. Using this method for both large-scale and small-scale projects, he always seeks to find the origin, the "true form" of the built reality by studying the territory, itself representing a set of elements working together according to the same necessity.

The reading of the territory can thus be conducted through the notion of organism, that is, of a set of territorial elements, settlement, structures, productive systems linked by relations of necessity changing over time. As in every built organism, it is accordingly possible to recognize phases and civil cycles that determine its formation, transformation, fragmentation and ruin.

He succeeds in giving an original interpretation in the reading methodology of a territory, often complex, of which he succeeded in showing unity. In other words, he manages to coordinate between the macro-scale and the micro-scale by also studying the territory's forming process and transformations.

Few scholars such as the German geographer M.R.G. Conzen or the Italian architect Saverio Muratori, from whom the logic of the architect's work seemed to originate, have investigated the laws of "forming process" in their studies. In fact, the Conzenian notion of process aims mostly to investigate the reality in reading the urban fabric of cities, whereas the Muratorian one perceives the built reality as a palimpsest, stating that "every form, however complex it may be, always bears every formative step imprinted on it, whether it is a gradual process or a synthesized one." Another scholar, the philosopher L. Pareyson, believes that considering the artwork in its formative aspects is to see it as a norm and as a guide, allowing us to understand its formation. All of these components seem to converge in Strappa's work, marking the relationship of necessity and understanding the site's character as a continuation of processes in time.

Two of his projects, similar yet different one from another, particularly describe the relationship between territory and architecture.

The first one, located in Milan, had for objective to link the different areas of the Expo. Dealing with formation by reorganization, the projects title "Expo after expo" clearly states the need to propose a macro-scale solution for the fragmented urban fabric of the city post Expo (Fig. 1.4). To do so, Strappa started by reading the territory's formative phases and ended up transforming the intersections of routes of adjacent centers into nodal urban places. This process, carried out through drawings, sketches and reading historical maps, allowed the design image to become precise and communicable.

Within the same project, two strong gestures reorganize the entire area and reinterpreted the urban fabric. According to Strappa's methodology used while designing, it is only after following the processual formation of the territory that the new typologies in connection with the existing can come to life.

Not everything that is historicized, anyway, has to be continued. The continuation of a formative processes is a critical action that involves the responsibility of the

Fig. 1.4 Study for the reuse of the Expo 2015 area in Milan, 2015

planner. Thus, while two small settlements, now confused in the melting pot of the Milanese suburbs, become new poles, their connection overlaps with the existing form, generating a sort of new linear district that involves the Expo area and gives new organicity to this part of the territory.

The second project tackles the relationship between territory and urban architecture considering the transformation of a vast area's formation by integration. Located in Mendoza, the proposal aims to regenerate a part of the city once occupied by the Trans-Andean Railway main station and now abandoned.

Starting by reading the formative phases of the territory, a clearer understanding on the intervention area and its lacking elements was possible. All the main axes forming the urban tissue led to the emergence of new functions between them. This created urban connections by stitching the different parts of the city and embracing the integrated parts of the urban organism.

The new design is therefore the latest stage in the transformation of an ongoing process, following a constant method in the production of the Roman architect.

Bibliographical Note

Barizza, E. (2021). *Louis Kahn in Rome and Venice.* Routledge, London.

Bettini, S. (1978). *Lo spazio architettonico da Roma a Bisanzio*, Dedalo, Bari.

Conzen, M. R. G. (1969). *Alnwick Northumberland. A study in town-plan analysis.* Institute of British Geographers, 1960, London.

Espuelas, F. (2012). *Madre Materia.* Marinotti, Milan.

G.R.A.U. (1981). *Isti mirant stella. Architetture 1964–1980*, Kappa, Rome.

Giannini, A. (1976). *Il tipo territoriale*, Istituto di Progettazione Architettonica, Genoa.

Heidegger, M. (1979). *L'arte e lo spazio*, Il Melangolo, Genoa.

Muratori, S. (1967). *Civiltà e territorio.* Centro Studi di Urbanistica, Rome.

Salamouni, R. (2023). Territory and urban heritage in Southern Arabia: The case of Al-Jahamah historical village in the Sarat Abidah Province, Asir. In *U+D, Urbanform and Design, International Journal on Urban Morphology*, n. 20.

Scardigno, N. (2018), *Landscape as forma mentis: Interpreting the integral dimension of anthropic space. Mongolia*, FrancoAngeli, Milan.

Strappa, G. (1991). *Metafisica e Architettura.* In VV. AA. *Un museo immaginario per Giorgio de Chirico*, Carte segrete, Rome.

Strappa, G. (Ed.). (2012). *Studi sulla periferia est di Roma.* FrancoAngeli, Milan.

Strappa, G. (2016). Nodi urbani. Contributi al dibattito sul dopo Expo di Milano. In *Paesaggio Urbano*, XXV, n. 3

Chapter 2
Method

Abstract At the beginning of the conversation, the problem of what could be the most suitable method for reading urban form in an operative way is examined. The problem is made complex by the current conditions of crisis that make a shared truth impossible. Against widespread relativism, it is argued that the architect should possess his own truth (even knowing that many others exist) that entails responsibility and allows choices. The proposed method is based on the notion of process, a circular procedure that starts from the built reality and, seeking the general in the particular, allows at defining a finite series of general principles (abstraction) from which to derive an infinite series of possible outcomes (action) in the production of the form. The process is not continuous, but even fractures should testify a procedural drama (an extreme, sudden transformation) inserted in the wide flow of built environment transformations. The form is not intended as the surface that we perceive of things: it is the way in which a structure is shown to our eyes, how we know, individually, even what we do not see directly.

Keywords Method · Circular process · Reading · Abstraction

N.S. I'd like to start our conversation by talking about the relationship between reading and designing, one of the central themes of your way of understanding the work of an architect.

In one of your writings you define the technique of reading as the "art of the possible."

By emphasizing this relationship, you seem to recognize in the use of reading the ability to reveal and interpret the aptitude for transformation of the built environment. In other words, the reading would coincide, somehow, with the project itself.

My question is the following: if the project does not derive from, but coincides with the reading, could it not constitute a limit on the designer's freedom, forcing him to follow a process that will end up imposing its choices?

Could you tell us more about this aspect of your working method, quite unusual in a context in which the center of the architectural project seems to be creativity and invention?

R. Salamouni et al., *Architecture in the Making*,
https://doi.org/10.1007/978-3-031-86833-7_2

In other words, how does the individual judgment of the designer intervene? That same judgment which in your method, in my understanding, tends to manifest itself as a "critical distance" from the reality that is offered to the designer.

G.S.

Let me begin with a brief premise. I believe that the architect, in the crisis conditions which we are experiencing, cannot but be biased and propose the truth he believes in. At the same time, he is aware that in the contemporary context, when the unity of things has been lost, this truth can only be partial and that there are many others.

In other words, the architect should possess his own vision of the built world in which he places and gives meaning to things. This vision should not only contain the perception of what surrounds us. It should also contain the myth and utopia that allows us to detach ourselves from the daily contingent and explore new territories.

However, within this idea of the world we inhabit, expressing ourselves based on a personal point of view is not enough.

We need to scrutinize with the reason the process by which the built landscape was formed and could be transformed. This could be done through a reading that will inevitably contain the result of a pre-formed vision. In fact, every reading in architecture is critical and contains the aptitude of the object to be interpreted. It also has the active intentionality of the reading subject. We must know that there are many truths, but that we can only own one.

Moreover, the partiality of which I am speaking is also concentration. While being conscious of the general context in which we operate, we should, critically, neglect some elements, so that others appear with the utmost clarity.

On the contrary, it seems to me that the architect nowadays is lost in an attempt to understand and share everything, ending up adapting to all the requests, which are those of the market. The proof is that today architecture criticism doesn't seem to exist except as an organization of consensus for media circuits. In fact, the general condition of architectural culture is still that of postmodern relativism. It is immersed in the late romantic myth of subjective expression and creativity.

Let me explain myself better.

Of course, the architect's task is to grasp the instances of his own time. This should not consist in "recording" the contemporary condition. Instead, it should be a biased oriented intelligence that seeks order in the built world by stitching together its scattered details according to its own unified design thinking, not embedded in the present.

Architecture understands the world by turning to the history and lives oriented to the future it designs. The present is the current condition that we need to change.

Many of the best-known works, which are supposed to represent the image of our time, seem to derive, rather, from an effort of mimesis, and are often a parody of the contemporary world. They constitute, in fact, the product of an acceptance proposed as inevitable, the adaptation to a condition of irresponsibility. Evidence of this disengagement is the obvious decadence of the monument theme, of the celebration of institutions as the result of shared civil values whose last true interpreter was the epic poetry of Louis Kahn.

But let's start, as you suggest, from the most personal and difficult truth to communicate. Let's talk about the identity between reading and design. I believe that things are exactly the way you put and describe them: this coincidence between two aspects of design work, which are generally identified as successive stages of a process, is a different way of knowing the built world. It tries to grasp from the object of study, not only the real characters, but, at the same time, the aptitude for its transformation. It contains a hope, a desire.

On the one hand, reading is, for the architect, a critical knowledge and cannot be neutral. It is purposive and contains an evaluation and a distinction. It does not precede the project; it builds it, informs it, indicates its possible future form. I understand the difficulty in accepting these considerations in a discipline in which the reading (historical, topographical, morphological) usually precedes the design act in an often-ritual way. In fact, reading the built landscape that surrounds us is not a description through which we state the terms of the problem. It is the substance of the project itself. On the other hand, the project is a general form of knowledge, a reading. As architects, we know the world around us through the project which, in some way, contains our universe, our way of being on Earth. It encloses a "pre-form." I believe that this notion explains many things: the pre-form is the condition within which we read the built environment, the substratum of our consciousness that makes every reading, and therefore every project, individual and non-repeatable.

This observation runs through the entire history of Western thought in various ways: from Aristotle's statement where the interpreter cannot "prescind from himself" up to the "pre-comprehension" **of** Hans Georg Gadamer (the a priori meaning we attribute to things when we need to recognize them).

Hence the architecture I am thinking of is a "formative" architecture, based on formed and forming. Therefore, the design activity is a circular process. It begins with reality: in the infinite possibilities of change in the existing reality, we look for some common aspects, the general in the particular. Through the study and comparison of the phenomena, we abstract some (few) generative rules. We trace a finite series of principles capable of generating an infinite series of forms (not all necessarily realized or achievable).

We then go back to reality, which only accepts those forms congruent with the real context, and whose number is further reduced when the responsibility of the designer intervenes with his understanding of economic and social needs.

Moreover, when he operates, the architect's aesthetic synthesis can only be personal, singular. Therefore, the general and common is "identified" and becomes unique, irreproducible and individual. In other words, I am convinced that our action should be placed in the great flow of transformations of the world we inhabit. We must seek the sympathy with things. In fact, for our small contribution to be useful to the transformation of the environment, it must contain a much distant and different way to design. This term has for me the literal meaning of inventing (from the Latin *invenire*, to meet), encountering new forms, in the study of the built universe, which may contain some fertile diversities or dissonances to cultivate.

Reading, as a critical representation of the built world, is also an individual attitude of the spirit. It is an art. As you said, it is the "art of the possible." Understanding it as an opening toward the infinite possibilities that are offered to the way we perceive the world implies a study of the form (a morphology) that is anything but classificatory. It should not be conducted with the tools of the entomologist, as if the universe built by man was governed by rigid and inescapable laws.

It is, on the contrary, a procedure that I would call "artistic."

According to a late Romantic idea still widely practiced today, is artistic the manifestation of a feeling as an individual. However, what I mean by art is, instead, in the classical sense, the rule that presides over the act of transforming things, the τέχνη.

The project consists, as I see it, in the last stage of a reading process, the one where what is ends and begins what could be and what you want it to be. It is evident that giving form to the aesthetic synthesis that transcends construction is a task that involves the architect's individual nature, his choices, and his poetic way (the personal truth of meanings) of seeing things. But it also has to do with his responsibility, setting boundaries to his artistic freedom.

In fact, we need to understand what this term means in architecture. The writer's freedom is not to use words freely, without respecting the structure of the language (without which there is no communication) but, on the contrary, to know it so well as not to feel its burden in writing. And the same applies to the musician. Why shouldn't it apply to construction?

Architecture, understood as a discipline, has its own character. It is a civil art, aimed at satisfying the need to represent a social context. It is an "inevitable" art: we may not see a painting we don't like but we cannot avoid the reality of an architecture, of an urban space. This is why we are not allowed to guess through perception: we are required to deduce through reason.

This way of seeing things gives a different meaning to some established axioms. On the basis of what I maintain, for example, the masterpiece is not the work that owes its value to the fact that it is totally new that it contrasts with what has already been produced. The masterpiece is the individual (individual!) synthesis of a forming process, the one that expresses, in the fullest form, the spirit of a civil community in time and space. It is formed through shared principles; the unitary spirit that binds the dispersed detail to the totality of the organism.

A masterpiece is the work (very rare) that looks at the past with new eyes, concludes a process, and sows the seed to start a new one.

N.S. There is one aspect that is still not clear to me. I would like to understand whether in your work the reading of the built reality offered to the designer represents a "limit" or a "boundary." That is, whether it constitutes an instrument of almost absolute value, by which the design outcome is strongly conditioned, or whether it is the center of a method that collaborates to the designer's work by defining its contours open to multiple interpretations, establishing a "critical distance" with the object of study.

G.S. I believe that the essence of the architect's work consists in the critical reading of the built reality open to different design outcomes and to infinite aesthetic syntheses. As architects, we operate on the existing built landscape through "pre-forms." They are different for each of one us and constitute the individual condition of our consciousness through which we look at the world.

Let me start with a consideration on the contemporary meaning of what I am saying.

In the crisis we are facing, I realized how statements like the ones I made may seem, in many ways, out of date. This consideration is very comforting to me because I almost agree with nothing in what is being done in architecture today, such as the aestheticizing drift in which even ethics has become an ingredient of the spectacle, of the media circus that turns everything into goods.

"More ethics and less aesthetic" is a successful slogan, fed to a society of consumers, in which no one really believes. I therefore accept the outdatedness of my research and consider it a resource. Although there is no doubt that it is a component of it, the essence of architecture is not a subjective expression, communication or personal commentary on reality: architecture *is reality*.

I don't know if this way of seeing things can be defined, as you said, as "critical distance." Certainly, by the very nature of our work, the relationship with the matter to be transformed tends to be passionate and emotional. But, according to a still very current romantic tradition, it threatens to overlap with the civil role that the architect has in our society.

The antidote against the intrusiveness of emotion is dialectics, which forces the thought to alienate itself from its own beliefs, to put a distance between the author and what he produces, allowing ideas, in a continuous becoming, not to stiffen.

In this sense, the built reality as well as the landscape it offers to our perception, do not constitute a limit at all. They do not dictate laws or rules. According to Kant, the limit presupposes "space that is beyond," which contains a whole world with its own order. However, the world around us is an open whole, where everything overlaps and coexists. Reading and designing, with the difficulties that this action entails, mean precisely this: unraveling the complexity of the whole by not isolating the part, understanding how the elements of the built landscape that surround us are complementary.

This allows us to not identify ourselves with the object of our study. I would say that the designer's effort should tend to look at the built world with the detachment that allows him or her to understand places as a common good, in their historical development, with a kind of "critical empathy."

This will allow us to overcome the pure perception of forms.

The form, which is at the heart of our craft, is not the surface that we grasp of things: it is the way in which a structure gives itself to our eyes, how we know, individually, even what we do not see directly. The seduction of appearance is one of the great dangers of our work. That white house we see is part of a group of houses. But the white color is the visible property that we all perceive through the optic nerve. However, the way in which it is part of a whole, an element of a society of houses (that is, the links that bind the objects), is not visible. It derives from experience, knowledge and reflection, that is, from the specific culture of the observing subject.

This necessary understanding of things should be clearly highlighted and does not necessarily mean, in the project, affinity: it can be translated into opposition, if circumstances require it.

But even fractures and tears, when they are real, express a procedural drama, the extreme of sudden transformation, not the exhibition of the author's feelings. I am thinking of the many centerless revolutions of contemporary production, the "discovery" of the theme of fragmentation of the contemporary city which has become a true literary genre.

At the opposite extreme, we find the absence of transformation, the identification of intention with what exists, itself part of a process. This is the most difficult reading. Perhaps the architect should learn that, at times, critically reading a place can also mean carrying out wise maintenance while basically leaving what has already its own function, beauty, meaning.

A few days ago, my friend Vincenzo Latina presented on Facebook the case of the splendid Baroque church of San Gennaro located in Capodimonte, Naples "*before and after the care of the Archistar Santiago Calatrava*" who turned it into a kind of seaside disco.

An extreme example, among many, which indicates how much architecture is a risky job, and often induces to the invasion of places.

Our Minister of Culture rushed to incautiously declare that "*this experience, which mixes contemporary and baroque art in an unprecedented way, will have to show us the way.*" Capodimonte church remains one of the many cases in which the collective feeling of inhabitants is trampled on, by the narcissism of the architect. How many buildings, historical and not, have in fact been turned into matter at the disposal of the designer's ego, receiving the concordant consent of the "critics?"

Was there really a need to invade the magical territory of the Domus Aurea in Rome by a spectacular up-to-date intervention? Wasn't the space in the *Sala Ottagona* evocative enough, with all due respect to the media success of the studio that designed the work?

Perhaps it is necessary to explain that restoration is already a project and a transformation. And that, sometimes, even less is needed. I think we should rediscover the fascination and usefulness of the gaze that contemplates the reading of things in which no aptitude for transformation is recognized. "Is it really necessary to build a new public granary"—wrote Confucius 2500 years ago—"when the old one is still good and the new one will cost people a lot of sweat?" That granary actually possessed, like all things, not only a utility, but also, a magnificence of its own. The problem was that many, noted the master, did not know how to read it.

Bibliographical Note

Bollati, R. (1980). *Metodo di lettura delle strutture urbane attraverso le fasi evolutive*. Istituto Universitario Statale di Architettura, Reggio Calabria.

Caniggia, G. (2003). Permanenze e mutazioni nel tipo edilizio e nei tessuti di Roma. In G. Strappa (ed.) (1989), *Tradizione e innovazione nell'architettura di Roma capitale. 1870–1930*, Kappa, Roma.

Gadamer, H. (2013). *Truth and method,* Bloomsbury, London.

Grassi, G. (1967). *La costruzione logica dell'architettura,* Marsilio, Padua.

Ieva M. (2018). *Architettura come lingua: Processo e progetto*, FrancoAngeli, Milan.

Jones, A.N., Larkham, P.J. (1991). A *glossary of urban form,* in Historical Geography Research Series, no. 26.

Milani, G. B. (1920). *L'ossatura murale,* C. Crudo & C., Cremona.

Monestiroli, A. (1976). *L'architettura della realtà,* Umberto Alemanni & C., Turin.

Muratori, S. (1946). Saggi di metodo nell'impostazione dello studio dell'architettura. In S. Muratori (1980), *Storia e critica dell'architettura contemporanea*, (G. Marinucci editor), Centro Studi di Storia Urbanistica, Rome.

Strappa, G. (1995). *Unità dell'organismo architettonico. Note sulla formazione e trasformazione dei caratteri degli edifici*, Dedalo, Bari.

Strappa, G., Ieva, M., Di Matteo, M.A. (2003). *La città come organismo. Lettura di Trani alle diverse scale*, Adda, Bari.

Strappa, G. (2018). Reading the built environment as a design method. In V. Oliveira (ed.), *Teaching urban morphology,* Springer, Cham.

Strappa, G. (2023). Urban morphology following the Muratorian tradition, In *"DeğişKent" Değişen Kent, Mekân ve Biçim, Türkiye Kentsel Morfoloji Araştırma Ağı II,* Kentsel Morfoloji Sempozyumu.

Strappa, G. (2023). The Italian school of process morphology. Roots, methods and future developments. The origins of the morphological studies in Italy. In *Saj, Serbian Architectural Journal,* n.15.

Whitehand, J. W. R. (1991). *The making of the urban landscape,* Institute of British Geographers Special Publication, no. 26, Blackwell, Oxford.

Whitehand, J. W. R. (2001). The physical form of cities: a historico-geographical approach, In R. Paddison (ed.), *Handbook of urban studies.* Sage.

Chapter 3
Form

Abstract This chapter addresses the problem of the definition of form in architecture, also in its relationship with the visual arts. The definition of form given by the philosopher Luigi Pareyson as "both concluded and open at the same time, in its non-finite completeness" and Noam Chomsky's definitions of language, as an unlimited set of sentences constructed from a finite number of phenomena, are examined. The problem of context is also explored not only as a physical and material place in which the form is meant to deal with, but also a cultural, immaterial condition which generates choices and programs. In particular, the landscape could be grasped not only in its aspects related to perception, but as a visible appearance of a material culture, a transforming territorial structure. Some examples of projects based on the study of form as successive stages of transformation of matter are provided.

Keywords Form · Context · Immaterial condition

***N.S. Let's talk about a concept that you often refer to: that of "formativeness." A term that Luigi Pareyson also adopts in the field of visual arts to bring the notion of aesthetics back into the wider structure of his theory on formation, transformation and interpretation of reality. It would be interesting to understand in which terms this principle is assumed within the framework of your interpretation of built landscape and also, within your design research*.**

G.S. Yes, indeed, Pareyson's ideas about the different modes of artistic production influenced my way of seeing things. I had read his book *Estetica. Teoria della Formatività* (Aesthetics. Theory of formativity) republished by Sansoni in 1960, when I still was at the university. However, I hadn't realized its usefulness for architects. Even then, I was skeptical of easy parallels with other disciplines, often misleading and casually used in architecture in a folkloric way. Some years ago, my friend Francesco Rispoli, knowing my work, suggested that I should reread it carefully. I owe him the discovery of a world of thoughts complementary to my own, which seemed to clarify some intuitions I had while designing. It is a fascinating and risky text, as it could lead to establishing direct links between architecture and art forms that, actually, have very different forming. Instead, we must make the effort not to

R. Salamouni et al., *Architecture in the Making*,
https://doi.org/10.1007/978-3-031-86833-7_3

get caught in the coils of an ingeniously and seductively constructed text, reading it also with the lightness that allows partisan synthesis, grasping the essence of the method it proposes, based on the ability to shape things through a dynamic process of transformations.

As it often happens when reading architecture, I found in his text what I was looking for, perhaps beyond the author's intentions. Of course, that's not the correct way to read scientific texts. However, in this way, I found new things and, at the same time, similar to those I knew through studies of urban morphology. After all, the deepest and sharpest thoughts exist before we can express them; they are somehow pre-formed, lying in the depths of consciousness. Reading them is sometimes the spark that sets in motion mechanisms waiting in our minds. Even theories, for us architects, are the coincidence of reading and designing: the reality of the text and our predisposition to accept and interpret its contents.

Pareyson's reading reinforced and supplemented my belief that the built world consists of a state of dynamic equilibrium of matter transformed into material, passing through different stages: from the generation and transformation over time to the decay ending with the unraveling of the original characters to the return to a state of second nature. That is often the beginning of a new process, the substrate of a regeneration.

The architectural work, therefore, coincides with the very process of its formation. It does not stop but finds its own temporary unity, completeness and the moment of balance in which "the form calms down and at the same time gathers." Pareyson's explanation of the form as a living organism, with its own life, seemed not very different from what I had written years earlier in my book *Unità dell'Organismo architettonico* (Unit of the Architectural Organism). However, the philosopher from Turin shifted the point of view. He defined the rules that govern it as "internal legality," a term that seemed bizarre and which instead, as the reading progressed, opened new perspectives.

Form itself is an organism, and as such formed and forming, with its own internal laws that bind the parts into unity. This definition takes on a particular meaning in architecture. It is anything but neutral. It indicates that buildings can be regarded as the result of a permanent process of transformation **of matter into material**, and then into elements, structures, systems that work together to form (give shape) and build reality at any scale: the shape of the city understood as an ongoing and constantly evolving process. According to my biased and oriented interpretation, the territory is an organism where each element (the land and its use, the routes, the settlements) binds together in a relationship of proportion and congruence. In opposition to the well-established literature that described its contemporary conditions as hopelessly incoherent and dispersed, we could recognize the territory's own formative capacities, its aptitude for constructing an ever new, albeit unstable, organicity.

Considering Pareyson's notion as a designer, and even forcing some aspects of it, landscape could be grasped not only in its aspects related to perception, but as a visible appearance of transforming territorial structures. I had been struck by a passage in his book in which form was defined as a process that, prior to the existence of the completed work, acts as an immaterial entity, as a law of organization. It seemed, to my pre-judiced eyes, not too different from the notion of the a priori, synthesis of Muratorian traditions (which, however, applies above all to base building), but also close to that of pre-form, which I was also developing for special building, for that part of the built landscape where a fundamental role is given to aesthetic synthesis. It also seemed to explain the distinction between "form" and "shape" employed by Louis Kahn. Pareyson's text highlights the logical and demonstrable aspect of principles that Khan had by intuition. I don't know how far these interpretations are a forcing: I believe that some sentences used by Pareyson, moreover, deliberately contains a part of polysense.

I find his definition of form as "both concluded and open at the same time, in its Infinite completeness" still perfect today. It contains the generative idea of type, the belief that a limited number of forms can contain unlimited individuations, inexhaustible ways through which the general becomes individual, uniquely unrepeatable. A designed and constructed form.

I have mentioned how the idea of generativity is important to the architect. I would add that the term is complementary to that of formativity. It is, again, a notion that arose within disciplines other than architecture, particularly in linguistics, and therefore to be considered with attention. Yet Noam Chomsky's definitions of language, as an unlimited set of sentences constructed from a finite number of phenomena, and that of grammar as a tool that enunciates its principles of construction, are certainly fertile notions for our work as well. What is particularly interesting is the corollary, which follows from these enunciations, i.e., the idea of "competence" understood as the ability of the speaker-listener to associate sounds and meanings according to the rules of his own language. It is a different and fruitful way of seeing the notion of "pertinence territory" as the set of cultural characters, including architecture, which belong to a group of individuals, to a society that recognizes itself in those characters.

Of course, the inertia of language, its resistance to change, is not the sometimes fragile one of architecture. Even in a world that we insist on defining as globalized, areal distinctions are evident, if we have the disposition to recognize them. As conducted in some current research in the Iberian countries, the permanence of a masonry plastic culture is an example.

It becomes more evident when compared to the typical experiments of the areas of Gothic traditions such as the Dutch one, based on light, transparent systems and structures with an elastic character of wooden derivation.

I think that all of my projects, to a greater or lesser extent, include this idea of generativity and formativeness. The one for new urban fabrics in Partinico, Sicily, for example, proposes to guide the generation of possible outcomes by establishing a law of transformation over time. It gives the opportunity to the inhabitants to transform their homes following the changes of the surrounding. In this case, the construction of the base buildings, made up of courtyard houses, had to take place in phases of organic growth, according to a forming process. It is based on four constructive stages, each aimed at satisfying, from time to time, the different needs of the inhabitants through successive densifications that would have also avoided further land consumption.

Fig. 3.1 Sketches for the "Silice e Fuoco" (Silica and Fire) exhibition at the Palazzo delle Esposizioni in Rome, 1992

Fig. 3.2 Photo of one of the halls of the "Silice e Fuoco" exhibition held at the Palazzo delle. Esposizioni in Rome, 1992

The setup of the exhibition *Silice e Fuoco* (Figs. 3.1 and 3.2), held in Rome in 1922 at the Palazzo delle Esposizioni and dedicated to glass art objects designed in the XIX and XX centuries, started from the idea of the same transparent container changing from time to time according to the works contained, while maintaining the same general "typical form."

N.S. I find it useful to consider the dialectic existing between the "designed form" and its context. A relationship that is, in turn, nurtured by a further relationship between the notions of "opposition" and "complementarity." These terms have to deal with "modification" and "continuity."

Could you go into the merits of these terminological dyad by describing the process of the ideational basis of urban projects that have been confronted with the "certain" context of the consolidated fabric, such as the projects for Carezzano, Piazza Mastai, Piazza San Cosimato, Fara Olivana?

G.S. Context is fundamental. It is not only a physical and material place in which the project is meant to deal with, but also a cultural, immaterial condition which generates choices and programs.

My method of investigation is linked to the historical context and the immaterial contemporary conditions of the area of belonging, understood in an open sense and willing to exchanges.

Let me explain it better.

When I define the fabric as a group of buildings that share some essential characters and possess common formative principles, I intend a biased reading referring to my context. In fact, "fabric" is not a universal definition. It works for a Roman building aggregate, where I live and which feeds my projects, and for those of many Mediterranean cities (Fig. 3.3). It also applies to different contexts, even to that of some historical Chinese fabrics, as some anthropic behaviors are common to all men, in any geographical context. But let's try to apply this definition to a set of single-family houses, jealously isolated within their own lot, of a North American suburb, which have no relationship with the neighbor's property except for a boundary, and we will considerable difficulties.

Reading is oriented, it is a cultural product of the reader. It happens for literature, visual arts and poetry whose meaning is the result, beyond the author's intentions, of the stratification of the many readings of unknown interpreters.

It also happens for the forms of architecture. The project arises as an interaction between the disposition of the object to be read, with its objective form to be interpreted, and the aptitude of the subject operating in its own cultural environment. Aptitude that depends on its own personal history made up of experiences, beliefs, values, emotions, which reads the object in ways that involve the individuality of the architect.

It seems to me that the examples you propose, testify in a different yet clear way to this method of proceeding.

The competition's call for the redevelopment of an area in the historic center of Carezzano, a town in southern Piedmont, implicitly indicated the demolition and reconstruction of some houses owned by the municipality in the historic fabric and the building of a new structure for public use (Fig. 3.12).

Instead, I believe that the goal of the new intervention should reinforce the character of solidarity that has been established over time between routes and public spaces. This has been determined by the nodalities of the routes, the relationships between housing, base building, and special building.

The new architecture was proposed as an urban node that transformed the old fabric into a public space.

The buildings to be transformed had features that clearly indicated the derivation from three original courtyard houses, according to a type also widespread in many other small Italian historic centers with consolidated rural traditions (Fig. 3.6).

Fig. 3.3 Proposal for the "Vivienda y Ciudad" competition dedicated to the renovation of the southern part of the Diagonal in Barcelona, 1989

The fundamental choice was to upgrade and update the existing fabric, consisting of buildings in a state of abandonment, trying, as far as possible, not to demolish anything.

To do so, a typical forming process of these constructions has been hypothesized through four phases of transformation (Fig. 3.13).

A first formative phase, characterized by the presence of a fabric of elementary courtyard houses, with access from the road.

A second formative phase, in which a partial infilling of some of the courts takes place, with the construction of smaller rural building volumes.

A third phase, in which some of the courtyards, originally property of a single owner, combined together to reach the multifamiliarization of the original fabric.

A fourth and last phase, in which the recast of the building cells takes place, with consequent specialization, internal reversal of the original external routes and internal knotting of the routes to form the new building. A typical forming process of a "palazzo" (Figs. 3.7 and 3.8).

While the first three phases correspond, with a good approximation, to what actually happened, the fourth and final formative phase never actually existed. It corresponds to the project, to the desired outcome of a process in the making. The new building had to have a representative character highlighted by the public space of the courtyard, where the paving expressed the hierarchy of the paths that intertwined in an "internal square" to form a small new civic pole (Figs. 3.14 and 3.16).

Here, the reading of the existing fabric was not a description. It was a finalized operation that contained the physical and historical context of the place, but also the immaterial context of the designer's culture. The last, and most important for us, of these transformations, recognizable as a specific character of architecture of the Italian city, was the collaboration (by recast, increase) of the housing construction that formed the specialized buildings. The notion of "palazzo" (palace) has been used as a synthesis of the amalgamation between different housing units. I believe that this critical reading corresponds, among the many possibilities, to the individual trend derived from the habit with historical fabrics, from the recognition of the congruence between the hypothesized process and the real process. But it is certainly not the only possible one. It is a choice that also gives innovation to the inherited building heritage, due precisely to partly individual choices (Fig. 3.17). The design method employed is a product of the designer's culture, of the "pre-form" with which the built reality is read.

The example of Carezzano poses the controversial problem of the use of morphological reading in designing, often the object of disapproval for a supposed determinism of the processual method. It seems clear to me that morphological culture is here used as a substrate with which there is a long-lasting "habit." As it would be impossible to write a literary text by reasoning on the rules of writing, or as a musical composition cannot be elaborated by reflecting on the musical rules, in the same way, for architecture, the moment of aesthetic synthesis proceeds with individual "spontaneous" tools. Therefore, the design act arises "naturally" within the principles assumed. Its outcome is open, unpredictable.

The palazzo theme, after all, has always been at the center of my work: first through a sentimental and an intuitive interpretation, then more and more consciously, through a method that starts from the logical and communicable reading of the existing building, then arriving to the aesthetic synthesis.

For example, the old project for the unfinished Ca' Venier, presented at the 1985 Venice Biennale (Figs. 3.4 and 3.5), is based on the theatrical representation of the Venetian palace. Around the great void of the hall the life of the building revolves and wraps itself. This space, a legacy of the open courtyard of the *domus*, suddenly opens up showing the poetic essence of the place, the vision of the mask and of death. They belong to the most consumed and vulgarized reading of the city, with quotations from other architectures. This project is the result of a suggestion and a collage, a kind of pop postcard.

In the project for Carezzano (but also Mesola, Fara Olivana (Fig. 3.18) and many others), the theme reappears in a more mature form: as a generative substrate, where the emotion, the poetry of the rural world which persists assuming new forms through the solidarity between old farmhouses, summarizes the history of the project at the end of phases of progressive awareness of the built reality in a specific cultural area.

Unexpectedly, despite a participation that we considered "demonstrative," the conclusion of the competition for Carezzano was positive. The jury evidently understood how the continuation of an ongoing process, the reuse of existing buildings, in

Fig. 3.4 Studies for the inner space of Ca' Venier dei Leoni, Venice Biennale, 1985

addition to protecting the character of the built environment, allowed the good use of available resources with respect to demolition and reconstruction. It also allowed for considerable passive energy savings due to the techniques adopted, the considerable thickness of the existing walls and their shape, and finally the location and exposure of the original buildings. Resources that it would be senseless to squander.

You've asked me what the problem of the relationship between continuity and modification is.

In my opinion, it constitutes the hidden theme in every project. It acquires great evidence in the relationship with the historical fabric of which Rome, with its complexity, constitutes the case of exemplary clarity.

I believe that the project for piazza Mastai (Fig. 3.24) can exemplify my idea of the transformation of a Roman public space. First, it must be considered that proposing a new space designed for Rome is not a simple project operation, at least in the sense commonly attributed to this term. It means thinking, I believe, of a condition of future ruin, of the image that the years will assign to the work that is being built today. We can foresee only a state of temporary passage, a transition. It is necessary to be very

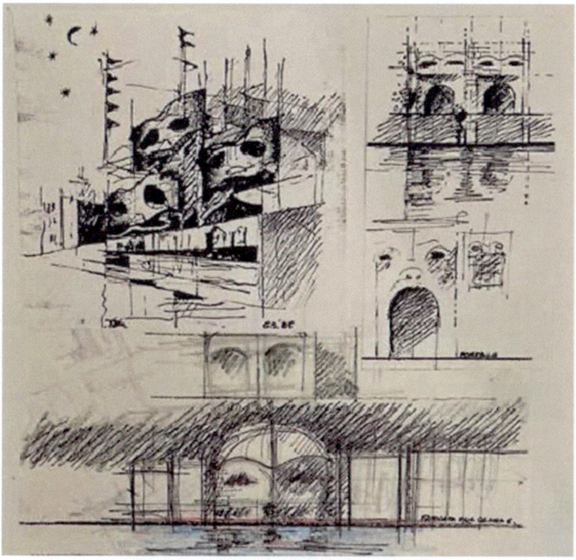

Fig. 3.5 Studies for the completion of Ca' Venier dei Leoni, Venice Biennale, 1985

Fig. 3.6 Competition for the redevelopment of an area in the historic center of Carezzano. New internal square. 2011

careful in planning these transformations, to curb the architect's propensity to invade places: the changes must be necessary, proportionate, congruent, often of modest entity.

The square then, particularly the Roman square, cannot be read as an urban void. It commonly faces this problem: a free space to fix. On the contrary, it is the result

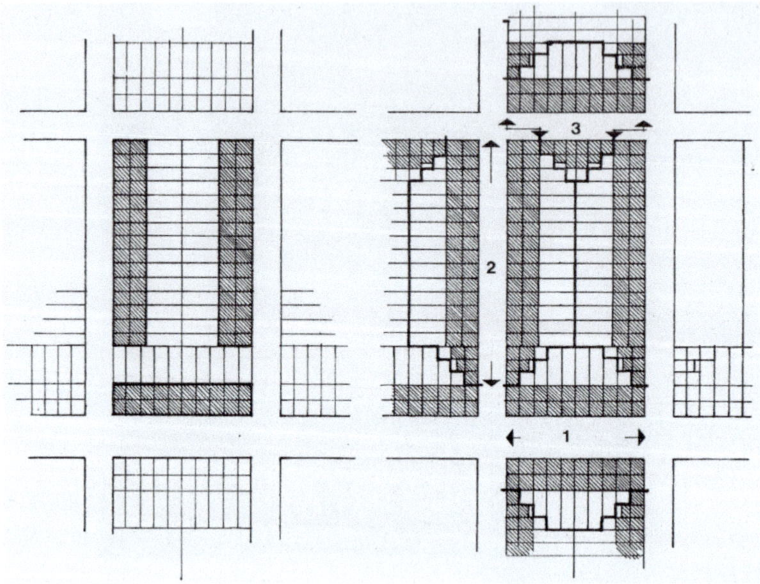

Fig. 3.7 On the left, formation of an urban block by progressive phases of construction. On the right, formation of the infilling fabrics in the pertinent areas of the row-houses on a matrix route (1), a building route (2) and a connection route (3) (from Caniggia 1979)

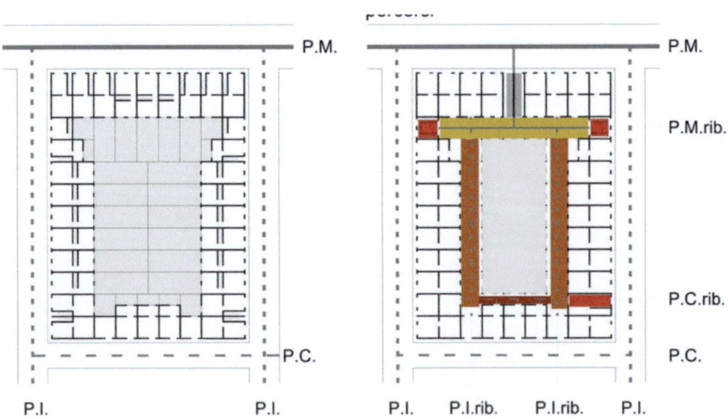

Fig. 3.8 Formative phases of special building by transformation of existing fabric (from the lectures of Giuseppe Strappa at Roma Tre Faculty of Architecture, AA 2022/23)

of a process, "filled" with all the intersections of forms that derive from its nodal condition.

The problem of the form of a public space can only arise here in the great stream of transformations that have built the current forms by updating the matter transmitted

Fig. 3.9 Sketch for the renovation project for Piazza Mastai in the Trastevere quarter, Rome, 1996

by history (Fig. 3.25). On the other hand, the square built by Pope Mastai did not respond, in my opinion, to any real historical urgency, since the place did not have a great polar vocation. On the contrary, it was the result of a highly imposed planned intervention whose conflict with the consolidated fabric is still clearly legible today.

In fact, in the XVIII century, the area was completely antinodal with respect to the structure being formed in Trastevere. It constituted the "back" of the building aggregates and specialized buildings that were forming (Fig. 3.9).

Therefore, the new design of the square domestics in part the pre-existing recklessly polar layout, centered on the fountain designed by Busiri Vici, interpreting its role as a vast space open to daily life, around the perimeter of which few functions have been placed, with a large part of the hemicycle enclosed by the blind wall of the Augustinian Convent (Fig. 3.26). The new space is dilated and axial: oriented.

Piazza Mastai, which was reduced to an unseemly parking lot, needed an "urban consolidation" intervention that would enhance its neoclassical vocation distinguishing it from the XIX century laceration of Viale Trastevere, and from its swirling river of traffic. Moreover, the resurfacing had to establish, through materials and design, a friendly relationship with the inherited spaces, to be flexible enough to qualify, while maintaining a recognizable design, the system of planned squares (Mastai, Sonnino, Cairoli) while adapting to changing local conditions.

The route of Via della Luce has thus been modified for the entire length of the ex Manifattura Tabacchi building (Fig. 3.27), adapting to the axiality of the new

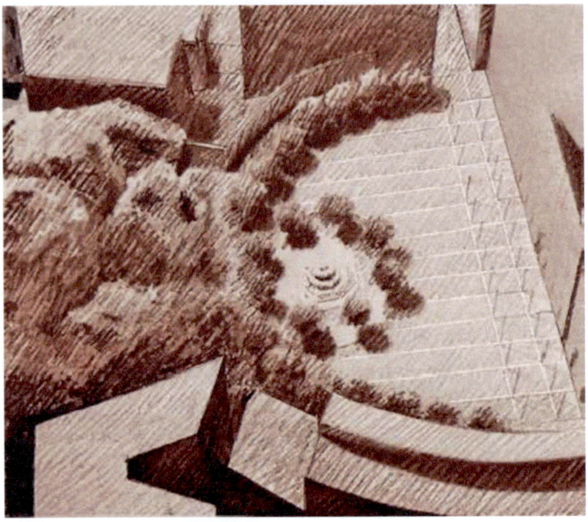

Fig. 3.10 Sketch for the renovation project of Piazza Mastai in the Trastevere quarte, Rome, 1996

square: the curvilinear shape, together with the increasing height toward the axis of the square, highlight the centrality of the main entrance with respect to the external space (Fig. 3.10). The serial character of the facade imposed by Sarti is thus forced, as anyone walking along this part of via della Luce can perceive today. This is the experimental (limited, due to the respect that the environment offers) and demonstrative dimension of the project.

Still in the historical environment of Trastevere, the project for Piazza di San Cosimato, which hasn't been realized, responds to the same principles. However, it considers the character of a large urban node of the place starting from the V century (Fig. 3.29). The route connected the nearby *posterula* of the city walls to the pole formed at the intersection between via Aurelia and via Settimiana, in correspondence with which the main polarity of the neighborhood then developed, with the formation of the convent of S. Callisto, of the basilica of S. Maria in Trastevere and its nearby square.

Without substantially modifying the layout of the place, the project proposed to clarify its latent form under the curtain of decay (Fig. 3.11). The arrangement of the XI century *prothyrum* area should have had a propitiatory value, indicating the beginning of the recovery of the entire complex of the Benedictine monastery of SS. Cosmas and Damian (Fig. 3.30). Of course, the theme of the square is very peculiar. It is a place where the memory and life of a neighborhood or an entire city is concentrated, an urban node in the broadest sense, which needs an aesthetic synthesis (Fig. 3.31).

For the recent project for Fara Olivana con Sola, a Lombard municipality made up of two nuclei now separated by a motorway (Figs. 3.19 and 3.20), the problem was to reconstruct a continuity. We tried to obtain it by redeveloping the squares with a common language (Figs. 3.21 and 3.23) based on the elementary forms of rural

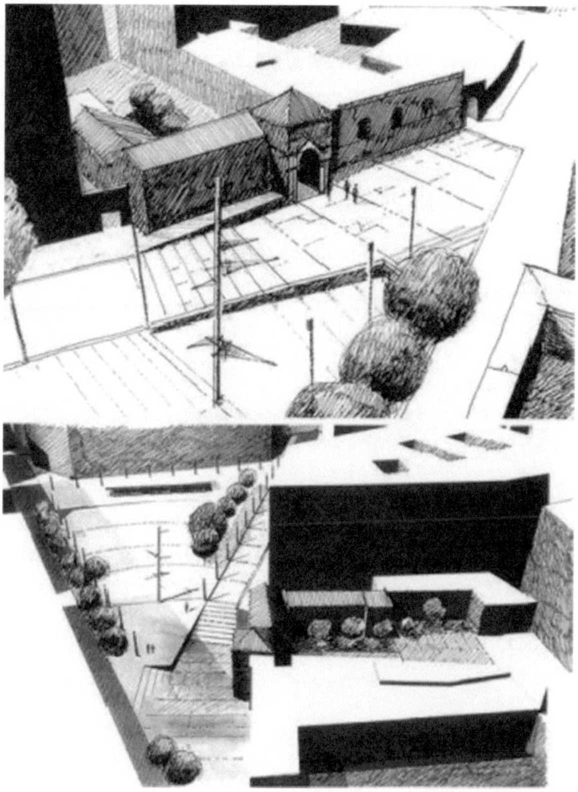

Fig. 3.11 Sketch for the renovation of Piazza San Cosimato in Rome, 2002

architecture and on a notion of fabric derived from the Roman *centuriation* that we have reconstructed.

One final observation. I believe that the same attention devoted to historical fabrics should be given to the suburbs forming process. In my studies on the historic center formative processes, as well as in the areas of the Roman suburbs that have been a place of wild speculation, I have tried to demonstrate that they constitute the same historical territory.

In fact, not recognizing the latent form under the concrete pouring that happened during the *boom edilizio* (building boom) in the sixties or the rule of a still possible organicity means affirming that here there is no order, and therefore everything is possible. And in fact, we continue until today, to intervene without any principle that gives meaning to building expansions.

Project A: Competition for the redevelopment of an area in the historic center of Carezzano Maggiore (Alessandria) for social and civic uses. 2011

(First prize winner)
G. Strappa (group leader), A. Camiz, M. Maretto, P. Carlotti, M. Longo, G. Galassi, P. Ciotoli, N. Boggio.

Historic fabric. *The central theme of the project was the revitalization of a portion of the historic fabric of Carezzano Maggiore, currently occupied by abandoned constructions, as an "update" of existing buildings. By building a small new urban center capable of unifying most of the public services in the settlement, the projected architectural intervention suggests the total reuse and partial transformation of the existing building, capable of responding to the functional program of the call as well as to "continue" the character of the existing architectural heritage.*

Method. *The main goal is to propose contemporary interventions in the historic center by reading, interpreting, and preserving the urban fabric as a formative matrix of the new organism* (Fig. 3.12). *For this purpose, laws and dynamics of transformation that have configured the structure of the historic center have to be considered, allowing the transformation of the urban fabric in continuity with the typological process in which the organism is inserted.*

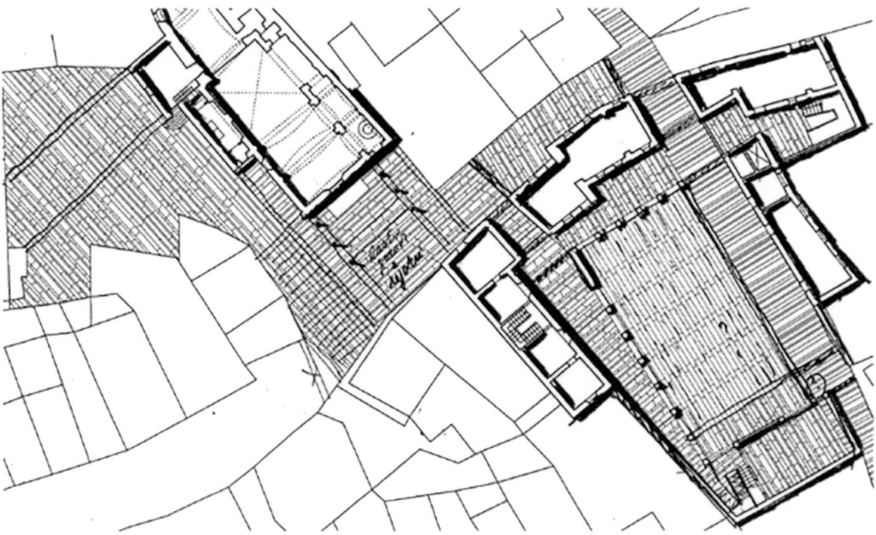

Fig. 3.12 Study sketch for the redevelopment of the Carezzano area

The forming process. *The design method consisted in critically reading the different formation phases of the site, understanding their transformations finally concluded by the new project (Fig.* 3.13*). We can schematize these phases through the following succession:*

*"It is the outcome of three intentions: the accurate reconstruction of the formation phases of Carezzano Maggiore, the contribution of that part to a new identity, the definition of a spatial warping which, through a careful and inspired decode of the metrics of the urban tissues, reorganizes the existing settlement enrolling it into a new set of urban relationships" (*Purini, 2014*).*

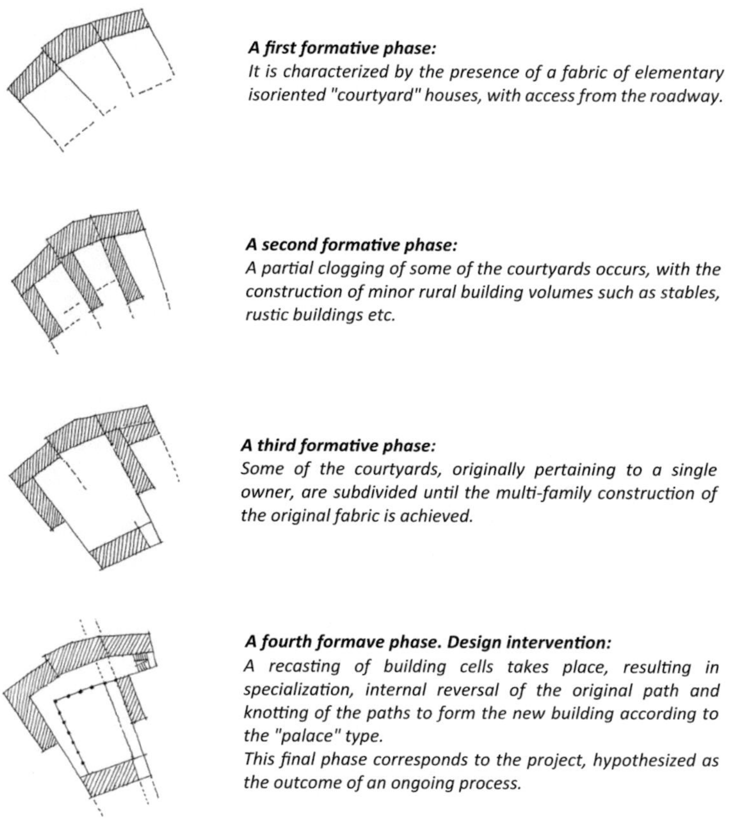

Fig. 3.13 Forming process of the building aggregation and the new design as last transformation phase

Fig. 3.14 General plan

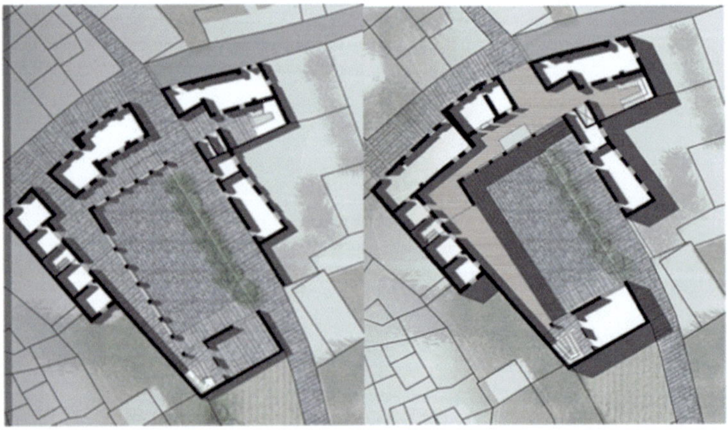

Fig. 3.15 Ground and first floor plans

Recasting process. *The proposed configuration is the outcome of a process of merging the three pre-existing courtyard houses in the assigned area, whose structures are unified in a public building and made organic by the critical synthesis of the design team. As in the formation of so many Italian palaces, the routes of the Carezzano settlement, which generated the historic building, are now turned inside to define the structure of a contemporary architectural organism. This mechanism generates the formation of a node that will represent the inner square of the new civic center, bringing together the pre-existing serial structures in an organic manner.*

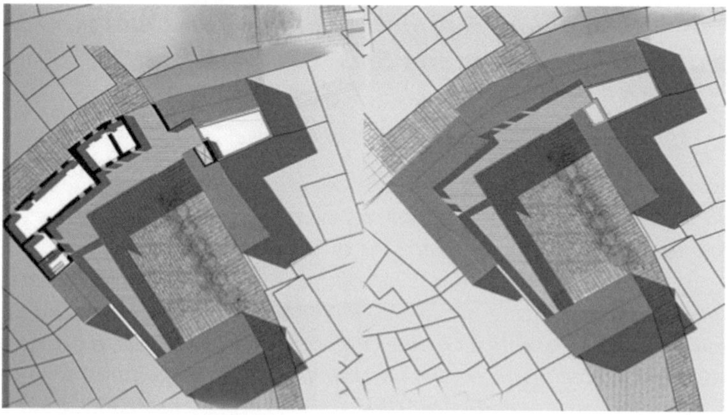

Fig. 3.16 Second floor and roof plans

Fig. 3.17 Project model

The project is based on the total reuse and partial transformation of the existing structures, which are reunified around an open space that constitutes an internal square as the true centerpiece of the new building (Figs. 3.15, 3.16, and 3.17).

Project B: Competition "A project to unite. Two squares, One community." Fara Olivana con Sola, (Bergamo). 2022

(Special mention)
G. Strappa (group leader), M. Ieva, F. Delia De Rosa.

The two ancient settlements of Fara Olivana and Sola are part of the same municipality but are currently separated by a motorway of national importance which has interrupted the territorial link between the two communities.

The central theme of the design competition was to collect and compare architectural ideas aiming to bring the two communities together by providing common spaces and a shared architectural image (Fig. 3.18). *The goal was to help in creating and strengthening a feeling of common belonging to the inhabitants.*

The required purposes can be traced to the following aspects: designing of new squares and multifunctional spaces for communities, renovating the outdoor areas of the existing urban services, retracing and rationalizing the road system of the various areas.

Fig. 3.18 Renovation and completion of existing rural buildings for the formation of a new square in Fara Olivana

Design of the new interventions as last phase of a territory forming process

The two small urban centers of Fara Olivana and Sola are part of the same community but are distant and separated by the new infrastructure. It is therefore important that the transformations include common signs rooted in the context (the strong plastic masonry character of the new buildings is linked to the language of the pre-existing ones) and that provide a shared and recognizable image to the collective spaces.

Character of the proposal*. The central theme of the project is the recognition of the identity traits of the architecture and the territory. A reading of the routes and ancient nucleus was carried out revealing some peculiarities, specific to the original formation. The study of a historical process of transformation of the territory and the settlements was fundamental. The intervention was intended to redevelop and update the existing fabric with partial demolitions (limited to indispensable cases).*

First formative phase (half of the XVII century):
Consolidation of the two settlements of Fara Olivina and Sola on the structure of the ancient centurial partition whose substratum conditions the new landform. There are no direct connections between the two centers, but routes that follow acommon centurial structure.

Second formative phase (contemporary period):
Contemporary situation. The territory is fragmented by the new roads, but the dimensions of the Roman subdivision appears as the character of the agricultural partition. In the settlements of Lombard origin remains the measure of the iugerum in the formation of fabrics of houses at court.

Third formative phase (new intervention):
The development of settlements follows the inherited form, proposing the formation of two nodal areas (also as support services to the Parco del Serio) complementary to each other as a bridge between the two communities. At the end of the two areas will be built two landmarks with similar architectural features.

Fig. 3.19 Territory's forming process and transformation

In addition to protecting the characteristics of the built environment, the design includes the reuse of the entire inherited buildings.

The territory and routes problem. *Taking note of the difficulty of connecting the two areas, also due to the presence of the motorway downgrading the zones of intervention, the proposal envisages some interventions that improve the perception of the Municipality as a unit. A new route intended for slow mobility tries to ensure accessibility toward the historic center. Parking spaces lost with the pedestrianization of the area in front of the church were created. Alternative road systems are strengthened.*

The new square of Sola. *The solution starts from the indication in the competition call to underline the character of the historic center of the town through the construction of a new town hall. A reading of the ancient nucleus* (Fig. 3.21) *was carried out which revealed some specific peculiarities of the original urban formation.*

Substrate module. *It has been noted that the structure of the present urban fabric conceals an older substrate module derived from the iugerum, typical forms of Roman planning in rural areas. Moving from this evidence, the project assumes the configuration of a piazza in front of the church S. Lorenzo. It becomes a multifunctional space,*

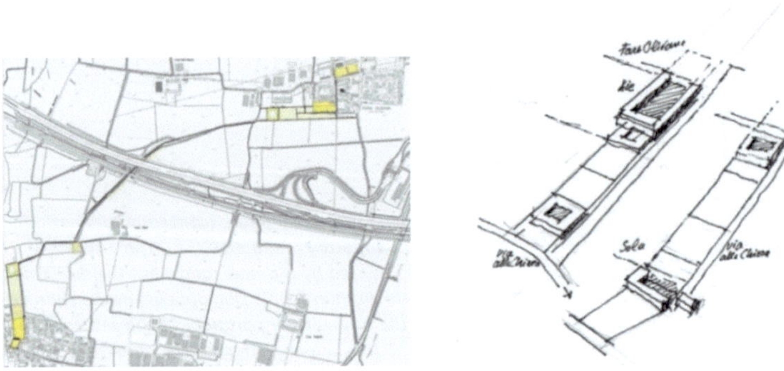

Fig. 3.20 Two areas of intervention cut by the motorway and the new expansions from Sola and Fara Olivana

Fig. 3.21 Design module derived from the juger (ancient Roman agricultural measure) partition substrata

Fig. 3.22 Nova sola square with the continuous façade connecting the pre-existing building and the new town hall

usable by the community and associations. It is opened to the outside and establishes a close "dialogue" with the existing, also through the continuous interactions that the building offers by projecting its own functions in public space. Attention was also paid to the relationship with the nodality of the church of S. Lorenzo which acts as the head of the square and, despite the continuity of the new building, is also visible from via S. Lorenzo through an opening in full height.

Continuity. *By joining the existing parish building, the new town hall forms a continuous façade that envelops the square, underlining its nodal function* (Fig. 3.22).

The design of the free areas along Via alla Chiesa has been redefined and directly connected to the town hall square through the which organically link the whole space. A place dedicated to children's play, useful, and adjacent to the nursery school, the tree system that surrounds the area and a large space dedicated to a flowery lawn, recover the typological mix of traditional Italian garden parks.

Fig. 3.23 Square in front of the S. Stefano Protomartire Church

The new square of Fara Olivana. *The space in front of the Church of S. Stefano Protomartire uses the same paving of the surrounding public areas* (Fig. 3.23). *No interventions are planned other than a small construction for the resting that defines the churchyard area. In the same settlement, the reuse of abandoned buildings in* Via *Vittorio Emanuele II and of the vast farm that delimits the southern part of the settlement, planned, to be used for fairs and public events.*

Project C: Renovation of Piazza Mastai, Trastevere quarter, Rome. 1996

(Competition winning project)
G.Strappa (group leader), A. Trombetta, F. Ricordy

The central theme. *The goal of the intervention was to renew the space of Piazza Mastai, distinguishing it from the XIX century laceration of Viale Trastevere, and its swirling river of traffic, through a new architectural design. The renovated square had to establish, through materials and design, a relationship of "sympathy" with the inherited buildings. Viability, landscaping, lighting, and paving are therefore the four main ideas of this project.*

The two main categories of works include first the repairing the damage suffered by the square due to its misuse and second the prevention of motor vehicles from entering the space of the square both to prevent new damage and to enable its actual public use (Fig. 3.24).

Fig. 3.24 The renovation of Piazza Mastai immediately after the completion of the works

Reading and Designing the Urban Fabric (Figs. 3.25, 3.26, 3.27, and 3.28):

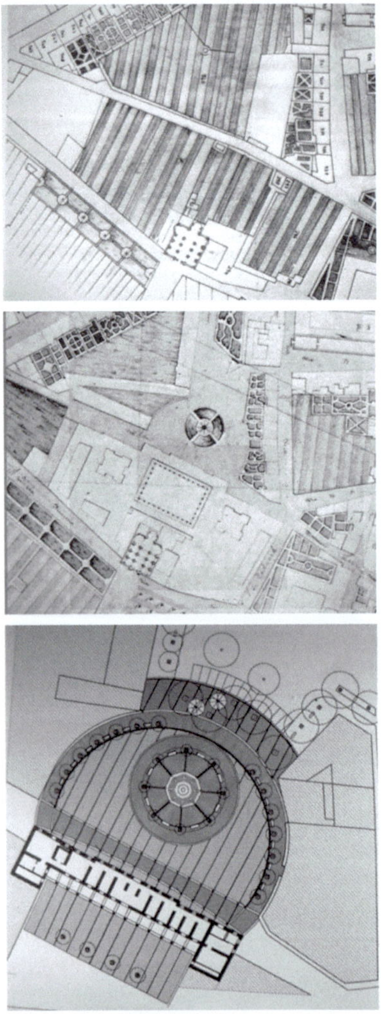

Cadastral map (1835):
The area consisted of equally divided land parcels between the matrix routes. The site had no special polar vocation.

Cadastral map (1864):
Square built during the Pius IX Mastai papacy and designed by Andrea Busiri Vici. The neoclassical Manifattura Tabacchi designed by Andrea Sarti was also built.
The circular shape of the square aims to establish a new centrality in the fabric of an area that is actually very marginal. Its highly imposed planned intervention was in conflict with the consolidated fabric.

Design Proposal (2000):
The new design of the square domestics in part the pre-existing recklessly polar layout, centered on the fountain de signed by Busiri Vici, interpreting its role as a vast space open to daily life, around the perimeter of which few functions have been placed, with a large hemicycle enclosed by the blind wall of the Augustinian Convent. The new space is dilated and axial: oriented (Fig.C.2, C.3, C.4, C.5).

Fig. 3.25 Forming process of the urban fabric and the new design as last transformation phase

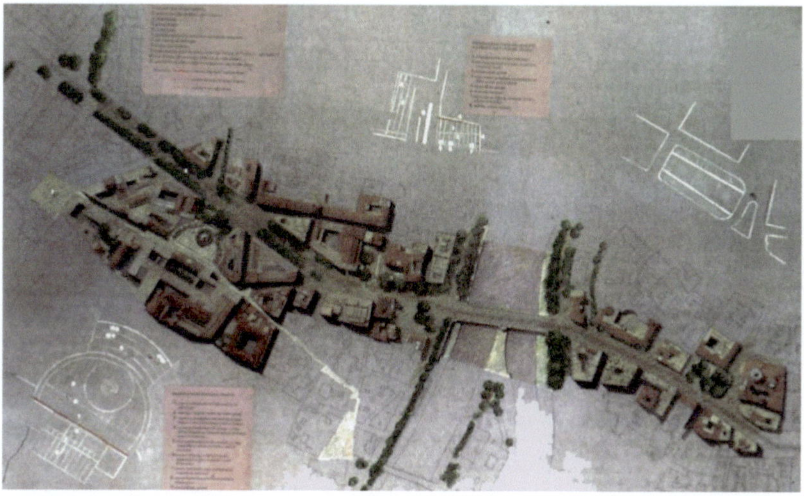

Fig. 3.26 Sketch of the project proposal in its urban context

Fig. 3.27 New Piazza Mastai square

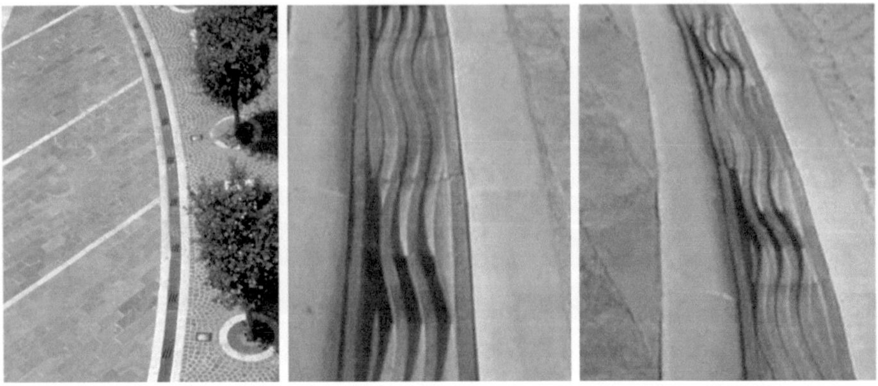

Fig. 3.28 Detail of the drainage system in cast iron

Project D: Competition for the Renewal of Piazza San Cosimato in Rome. 2002

G. Strappa (group leader), M.L. Tavella (designer), G. Carbonara (consultant), R. Bellardi (accountant), E. Fortunati (collaborator).

The central theme. *The objective of the project is to provide a new structure to one of the liveliest squares in the district, and an important local marketplace. The square is closely connected to the church of S. Cosimato, built in the X century, whose vast convent was transformed into a hospital* (Fig. 3.29).

The sense of the place. *Piazza San Cosimato could be considered as an urban node formed by two routes coming from the S. Maria urban pole, the first, and from the "foothills route" (of which the current Via Roma Libera reconstructs the middle section) close to the slopes of the hill of the ancient Horti Getae the second, routes that joined at the gap through the walls. The map of Rome by Mario Cartaro (1576) illustrates in an exemplary way the origin and first development of the triangular shape of the square as a masonry perimeter of the orchards and gardens at the intersection between roads.*

Continuity. *The main goal of this proposal is to preserve the character of the area as it has historically been determined through a very clear formative process well illustrated by cartographic and archeological documents.*

Fig. 3.29 New arrangement of the San Cosimato prothyrum

Forming Process of Piazza San Cosimato Urban Fabric (Fig. 3.30)

Project proposal. Piazza S. Cosimato can be considered as part of a more general system of pedestrian use of Trastevere's historic routes, polarized by spaces that are already becoming more specialized in function of a more rational use of the neighborhood's public areas. In this sense, a "quadrangular system of squares" can be identified, consisting of the squares of S.Maria in Trastevere, Piazza S.Cosimato, Piazza Tavani Arcuati, and Piazza Mastai.

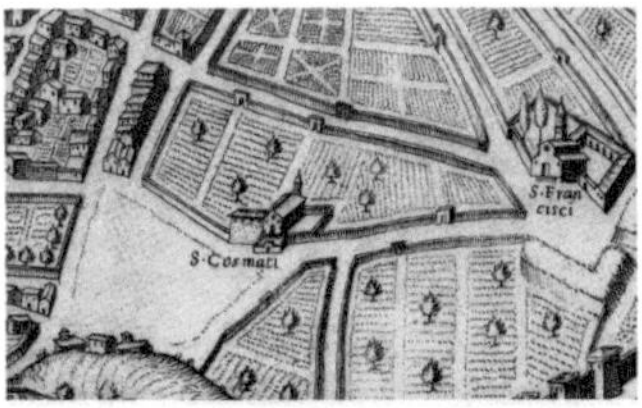

First formative phase**:** ***X-XVI century***
Origin and early development of S. Cosimato church and convent in a side of the triangular shaped square at the intersection of ancient routes that intertwined in the area of the ancient Mica Aurea. (Cartaro map, 1576)

Second formative phase**:** ***XVII-XVIII century***
Hierarchical structuring established by the growing urban fabric.
The XVII-XVIII century layout of the area confirms the formation of the triangular space indicating its vocation as a public space, to which the now almost final arrangement of the complex development of the Monastery of St. Cosimato contributes. (Nolli plan, 1748)

Third formative phase: XIX – XXI century
Completion and consolidation of the urban fabric around the square consisting mostly of multifamily "in linea" houses with high population density
With the construction of the new restructuring axis of Viale del Re (now Viale Trastevere),formation of a new system of urban polarities that give a new centrality to Piazza San Cosimato
Map of Rome and suburbs, 1924. IGM (Istituto Geografico Militare)

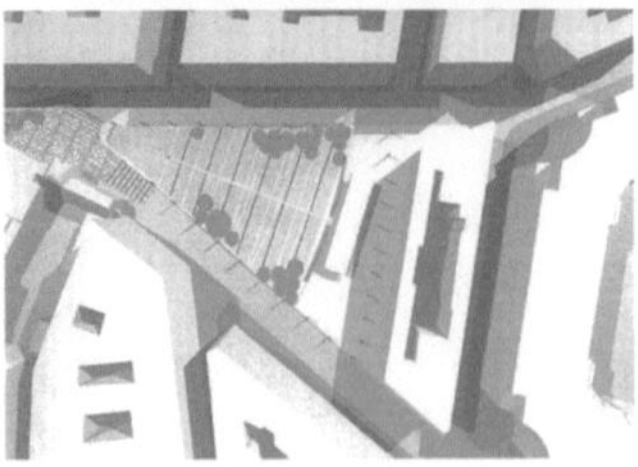

Fourth contemporary formative phase: new intervention
Consolidation of the triangular shape of the square. Extend, within the limits of what is rationally possible, the pedestrianization of the square and of the routes around it, indicating the possibility of closing to traffic, gradually and by time slots, part of the streets currently invaded by cars. formation of a more general system of public spaces aimed at upgrading the entire neighborhood.

Fig. 3.30 Forming process of the urban fabric and the new design as last transformation phase

***Setting of the monumental area (area A)**. The space object of competition inside the monumental complex of S. Cosimato and in front of the church of the same name, as indicated by the plan of the monastery of 1875, was built only on the prothyrum side, with a modest internal appendix. Thus, the spirit of this space is that of an open and protected place, which should be freed from the building additions (moreover, not significant) of the last century. Therefore, the paving will be made of different materials so as to evoke the character of places formed through reuse, syncretism, and recasts. Basalt paving stones of different types are used, in order to establish a relationship of affinity with the pre-existing, while avoiding imitation of historical forms.*

***Planning of central pedestrian areas (area B).** The central pedestrian area will be regularized to make the character of the place immediately legible. This regularity will be further accentuated by the formation of a central axis that will organize the geometry of the paving and travertine courses.*

The result will be a moderately concave and welcoming space, also suitable for representations and performances, but where, above all, the inhabitants of the neighborhood will be able to pause by sitting on the steps or on the double seating that closes this area, separating it from the market area (Fig. 3.31).

***Landscape elements**. The few existing trees will be supplemented by a large number of new plantings, consisting of plane trees and, especially elms, arranged in rows. Children's play equipment will be continuously replaceable over time. They may also*

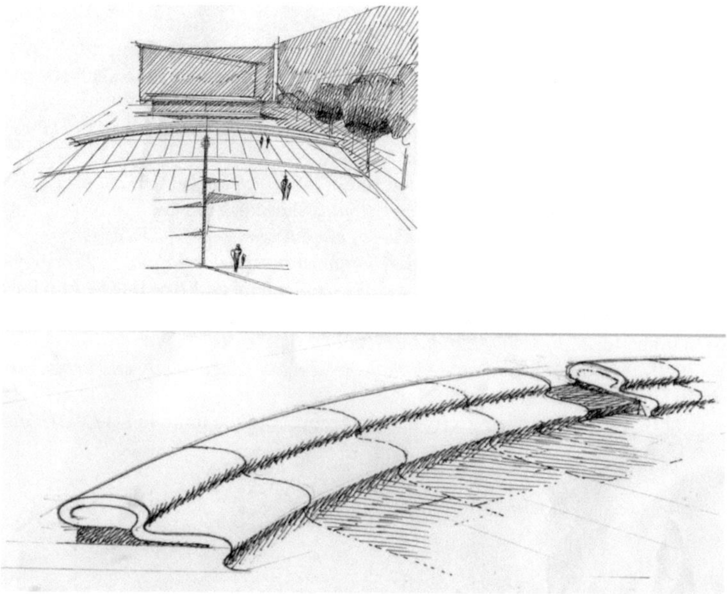

Fig. 3.31 Study sketches for new travertine benches

be placed near the market areas, but preferably in the southern area of the square bordered by travertine seating.

Market area (area C)*. Maintaining the character of a light and removable structure, the project proposes to gather all the fixed structures planned for the market under a single metal roof in the form of a simple barrel vault with very low curvature. This structure, located on the axis around which the public spaces are organized, will also constitute the functional polarization of the area reserved for the market. It will be dedicated exclusively to commercial functions. The roof will be made of copper-clad steel, the supporting structures of steel, and the different component elements will be made of painted aluminum.*

The Prothyrum*. The southern part of the square, due to its elevation above the area in front of the prothyrum, will form a "loggia" completely secluded from the surrounding pedestrian and vehicular traffic. This sector of the central area will be provided with continuous seating on the perimeter to accentuate its character as a secluded and protected space.*

Bibliographical Note

Buongiorno, V. (2020). *Suburban retail spaces. Formative and transformative process,* Springer, Cham.

Caniggia, G., Maffei, G. L. (1979). *Composizione architettonica e tipologia edilizia 1. Lettura dell'edilizia di base*, Marsilio, Padua.

Conzen, M. R. G. (2004). *Thinking about urban form,* Peter Lang, Bern.

Chomsky, N. (2006). *Language and mind,* Cambridge University Press, Cambridge (MA)

Eskidemir, K., Kubat, A. S. (2019). Urban form and culture: A comparative analysis of Anatolian and Italian towns. In *Urban Morphology,* Vol. 24, n.1.

Kropf, K. (1993). *The definition of built form in urban morphology.* Unpublished PhD thesis, School of Geography, University of Birmingham.

Kropf, K. (2018). Plots, property and behavior. In *Urban Morphology,* Vol. 22, n.1.

Pareyson, L. (1974). *Estetica. Teoria della formatività*, Sansoni, Florence.

Purini, F. (2014). *Designing a manifesto.* In *Paesaggio Urbano*, Vol. XXIII, n. 1.

Strappa, G. (2016). *City as a process: Rome urban form in transformation.* In *Proceedings of the 22nd ISUF International Conference*, Rome, 22–26 September 2016, U+D edition, Rome.

Strappa, G. (2022). The built lot and the future city. The elementary unity problem in the reading and design urban form, In *U+D, Urbanform and Design,* n.16.

Strappa, G. (2020). *Assemblage and aggregation: Reading the ancient city and urban composition methods.* In *Urban Morphology,* Vol. 24, n.2.

Strappa, G. (2020). The post pandemic city and the recovery of the limit. in *U+D, Urban Form and Design,* n.13.

Chapter 4
Organism

Abstract This chapter examines a contemporary notion of organism useful for architectural design. The new definition of organism and organicity, have little to do with the naturalistic matrices used through the imitation of nature. They rather express the aspiration, often disappointed but always persistently sought, for a unifying synthesis, which runs through even the most critical phases of architectural history. The idea of organic solidarity between the components that determine the appearance of the built form still continues in the center of modernity, like in much of modern Mediterranean architecture, using load-bearing and enclosing forms, organic binding construction, space, language in an aesthetic synthesis. The problem of designing organically, following the formative process of built reality, clashes today with design techniques undergoing a progressive phenomenon of abstraction. The media and manipulated interpretation of the built form is superimposed on the direct and natural perception of reality. Examples of design practice are proposed.

Keywords Architectural organism · Built form · Plastic construction

R.S. I would like to talk about the notion of organism, a term which was widely investigated by the Roman School of Architecture in the interwar period and then by scholars of urban form such as Saverio Muratori, Aldo Rossi, Gianfranco Caniggia, as well as by philosophers of aesthetics. More generally, it has spanned the entire history of modern Italian architecture, from the Renaissance onward. Do you think that this notion, in a different and updated form, is useful for the contemporary architect?

The notion of organism has always been equally useful and tricky at the same time. It has been, and is being, confused with the imitation of the human body as well animal or vegetal organisms. The human body itself has been among the most frequently proposed examples to explain the idea of organicity in architecture where, like a living body, each part should constitute an organ that performs a particular function and is necessary to the whole.

Perhaps the statements of past theorists have been interpreted too literally. In fact, no architectural organism has ever seriously built based on the forms of living beings,

R. Salamouni et al., *Architecture in the Making*,
https://doi.org/10.1007/978-3-031-86833-7_4

obeying zoomorphic or phytomorphic rules and suggestions. The famous drawing that Baldassarre Peruzzi allegedly used for Rocca Sinibalda Castle in the form of an eagle (the Habsburg eagle granted by Charles V to the Cesarini family) turned out to be a survey stored in the Uffizi in Florence.

This analogy has generated endless misunderstandings and false perspectives. The flying machine invented by Leonardo by observing the birds flying was a powerful poietic insight, but man has never flown with a flapping-wing device. Only later, by breaking down, through reason, the forces exerted by animal flight into the vertical vectors of lift due to the shape of the wing, and the horizontal vector of thrust provided by the propeller, did man produce the flying machine, which has nothing to do with imitating nature.

The definition of organism we are interested in, and the definition of organicity derived from it, thus have little to do with the naturalistic matrices used through the imitation of plant and animal forms. But I believe the analogy proposed by René Descartes between machine and living organism (which would also be a more complex machine) is likewise misleading. A machine that "having come out of the hands of God, is without comparison better ordered and possesses more admirable movements than those that can be invented by men," as he writes in his *Discourse on Method*. It had not been grasped, one would say, that the "forming" capacity recognized by Immanuel Kant and taken up by Luigi Pareyson that possesses every living organism, in which the individual elements are not simply assembled, according to a finalization, to form the whole, but are themselves conformed by the whole.

It is the same difference, after all, that passes between the Corbusierian conception of the dwelling understood as a machine for inhabiting, where the construction is an organism by the mere subordination of the parts to its function. Also, the house is understood as the result of a formative process in which the part adjusts, proportionally, and updates with respect to the unity of the whole through successive stages that become part, today critically, of the builder's consciousness.

In other words, what I am talking about is a contemporary meaning of a term employed in the past for very different purposes. On closer examination, the very etymology of the term "organism," in the sense we attribute to it in studies of the formative processes of built reality, relates to the more recent term, unknown before the Enlightenment, of "organization" as the law that presides over the coordination of elements with one another. The term was employed in Dante's time as well, but it had the literal meaning of "forming organs." It later entered scientific language by taking the value of "ordering, arranging." In the mid-XVII century, the physiologist Marcello Malpighi defined organization in a modern way as the term for the set of parts cooperating in the same function.

I think that the idea of organism useful to the architect is associated with entirely artificial structures organized in such a way that each part has an organic relationship, precisely, with the others, that is proportionate to and congruent with the rest of the structure. To simplify, I would say that organism meant, modernly, an "organization" that transforms over time, renews itself, adapts to new conditions.

The actual form of any architectural organism, from the building to urban aggregates, to the city, to the territory itself, should be considered as a special case among the infinite possible states within the great flux of matter transformation. A condition of temporary equilibrium of elements, linked together by time-varying and ever-changing nexuses that can be summarized, simplifying, by the notion of *motion* and the closely related notion of *process*.

The same forms that history has brought down to us would be only an expanse of muted fragments without the sequences of transformations that explain them, that gather them up and give them order, without the synthesis that in certain stages of civic maturity unifies in a few constructive gestures the complexity of developed experience and its memory.

The notion of organism expresses precisely the aspiration, often disappointed but always stubbornly sought, for this unifying synthesis, which runs through even the most critical phases of architecture.

Therefore, for an architect, the architectural organism is not a simple object, but the set, relatively autonomous, of elements linked by relations of necessity that unitedly contribute to the same end. It is the outcome and memory of a process that involves, beginning with matter, phases of transformation and successive "changes of state."

The *formation phase* is understood as conception of a structure in which the creator is not alone in facing the complexity of the problems confronting him but starts from historicized collective experiences.

The *construction phase* is the concrete change of matter transformed into material and then, by successive degrees, into a complete organism.

The *transformation phase* is a continuous process of modification and updating in relation to the conditions of the civil context.

The *phase of ruin*, finally, is an extreme change of state and return, in an inverse process to that of construction, to disaggregated structures (i.e., which have lost their internal bonds of necessity), then to elements, to material, until they return to matter.

This last phase of progressive return to the state of nature, which man tries to ward off through operations of deferment (restoration, rehabilitation, reuse, maintenance) never considered and foreseen in the project nor thoroughly studied, is the one that makes the character of the architectural organism most clearly recognizable. It testifies to its durability (the plastic capacity to "processively" ruin, generating the fascination of new, mysterious forms) or fragility (the sudden end, which leaves few, uncertain traces).

R.S. The center of your method in reading built reality seems to be based on transformation. It seems to me, then, that a fundamental role, in defining the form of the architectural organism, should be played by the character of the matter that originates it and the technique that produces it. I believe that this is an important aspect of your discourse and a very different idea from what is argued in design manuals, for which material and technique appear in the final stage, as tools for the realization of a design idea.

Certainly. Matter, such as material, and the technique that transforms them, are necessarily part of architectural invention; they are at the origin of the transformation process. And the origin is not only the moment in which things are generated, it is also their explanation.

According to a literal interpretation of the term (from Latin *trans - formare*, to change of form), the notion of "transformation" includes those of plurality and judgment. Built reality changes according to different directions, contained in the very essence of the thing to be transformed (object), among which man (subject) chooses according to his own critical choice. Transformation is, in this sense, the relationship that is established between the possibilities of the object to be changed, and the capabilities, or aptitude of the subject to change it. This subject is the *artifex*: artisan, artist, maker, architect. The object is the very substance of architecture: *materia signata* (marked matter, to use an expression of Saint Tommaso d'Acquino) that gives rise to different forms of inhabited space.

Historically, man has marked matter through pairs of opposite and complementary terms (dyads) "elastic" and "plastic" by distinguishing the link between matter and form over time.

It is a division that provides a synthetic orientation of the built world and runs throughout different architectural theories. It played a central role in European treatises, particularly XIX-century ones, until, at least, the elastic character of the elements of structures, of the systems that make up buildings, was the only one facing the future. Since then, serial and elastic structures have taken over as the authentic character of modernity.

This dyad distinguished in the past the specificities of different "cultural areas," meaning with this term the territories within which a large number of common features were recognized (not only in materials) at all scales.

The initial phase of the project, in architecture, began precisely from this recognition of the character of matter, of its aptitude to be transformed into material suitable for construction.

It seems to me that this dyad does not contain any determinism (in areas of elastic character, stone structures, not elastic at all, are actually used) so much so that the terms have over time taken on almost the meaning of metaphor. On the one hand, the world of elements with a serial vocation that form non-closing, load-bearing structures, coinciding in Europe with the Gothic area; on the other hand, the elements that give rise to load-bearing and closing spaces structures with an organic vocation, coinciding with the Romanesque areas, where transformations are slower, traumas are slowly reabsorbed and permanence's are more evident.

The term plastic thus serves to concisely express the idea of an architectural culture in the making, with an organic character, where the organic solidarity between the parts ensures that every modification remains imprinted and legible at different scales, and the existing matter is permanently transformed by reabsorbing even ruptures and discontinuities. A term that also indicates, implicitly the character of resilience of these structures.

Indeed, the organic construction possesses, at all scales, an organized structure capable of absorbing trauma by deforming and self-regenerating over time. A structure that needs a continuous regeneration to survive.

This notion of solidarity between the components that determine the manifestation of the built form still endures in the center of modernity. Not only, in architecture, in masonry construction, layout and expression were organically linked, but also in modern Mediterranean architectures built in reinforced concrete, load-bearing and enclosing forms are employed binding construction, space and its representation. See the cases of the works of Dimitris Pikionis, Sedad Hakki Eldem, Fernand Pouillon and many others. It is plastic architecture not because of the material used, but because of the organic essence that in-forms it: a plastic building culture in which the urban form and that of the territory are obtained by successive mutations that take place continuously over time from an initial material endowed with its own ductility. This is the case, precisely, of the Mediterranean city of ancient origin, where the routes, layouts, fabrics, are modified over time, always reproposing some characters of the previous phase of transformation.

At the opposite extreme is the city of eternal reconstruction, which proceeds by rapid changes, where "all that is solid melts into air" (title of a remarkable book by Marshall Berman) as in the North American metropolis of conflicts and competition of buildings for survival.

This interpretation allows even the modern inventions to be placed within the great flow of the built environment historical transformations placing even innovations such as the Lecorbusian free plan or the free façade within the transformation process of Gothic organisms based on point elements that carry loads but do not close spaces.

But, as is also always the case in architecture, the areas of greatest economic and political development impose their values. Henry Russell Hitchcock and Philip Johnson, in their *The International Style*, clearly indicate how elastic, linear, light, mass-produced elements express in the fullest way the character, linked to the machine and the world of industry, of modern construction. This is a true ideology that has nothing to do with technical performance in the use of materials.

They were placed within a world that had its own culture of established Gothic traditions. Clearly, this is a biased interpretation of built reality and its possible future.

The rediscovery of the modernity of the plastic world, therefore, of its reading and its topicality for contemporary design, is, as you can see, a very fertile theme. And it is a theme that has always interested me. I have always tried to give my projects a plastic and massive orientation. I realize that now. It was a spontaneous choice, not consequent to theoretical reflection. After all, if one has to write a story or compose a musical score, one cannot think about the rules of writing words or music. Choices emerge from a cultural substratum deposited through the rational and transmissible reading of the constructed and its critique. Therefore, our faculties should teach how to read the built world not only through perception.

As far as my personal experience is concerned, I have not received, in the course of my studies, any guidance on these issues. Their discovery was a difficult achievement and, in the end, their concreteness almost a revelation.

It was a most useful exercise to have built, as soon as I graduated, the row-houses in Via Dalbono, Rome. Even my way of drawing changed a lot after this experience (Fig. 4.1).

R.S There is a wide number of prejudices today against the notion of organism. One of the most widespread is that it has no value for interpreting the architecture of the past but has entered a phase of irreversible decay in the transition to the modernity. The new technologies and problems posed to the architect, always requiring new solutions, should, according to some designers, have made it unnecessary to study, in the contemporary condition, how the built reality was formed on the

Fig. 4.1 Detail of row-house in via Dalbono, Rome (1973), and the new porch of the Terni cemetery expansion, Terni (2010)

basis of organic relationships between parts of buildings, urban fabrics, cities, and territories. Instead, you seem convinced that the crisis into which the notion of organism entered with the advent of the modernity should be considered in the same way as all the other successive crises throughout the history of architecture: a phase of rapid transition in which many certainties collapsed, and as yet unknown problems were faced. Thus, a period of fertile transformation.

Yes, it is certainly true that one of the main reasons for perplexity on the part of extensive sectors of the discipline, and a reason for frequent criticism of the processual method, is its applicability to contemporary reality. While it is recognized that there is a permanence of features in pre-modern fabrics (within certain cultural areas and historical phases) such that it is possible to speak of organisms in transformation, it seems hard to grasp with similarly synthetic tools the modern modifications, still taking place, that determine the present form of the city. Modifications whose understanding is, moreover, indispensable for contemporary design.

For the present-day city, the only usable notion seems to be that of "complexity." But this notion is now, in architecture, equivocal as it has become, in usage, all-encompassing. It seems to document, but not explain, the contemporary fragmentation of the formative unity of the city's architecture.

But let us start from the root of the problem.

In fact, design techniques have undergone a progressive phenomenon of abstraction over the past two centuries. The media and manipulated interpretation of the world is superimposed on the direct and natural perception of reality, making it possible to "liberate" form from its concrete bed, to detach it from the ground. The architect can propose forms analogous to those of the real world read in its disintegration, to pose problems traditionally belonging to other disciplines that operate by right on form detached from reality to the domain of the visual arts, forgetting that architecture is not description of the reality. *It is reality*. Thus, forms are separated, in space and time, from their natural surroundings, severing their reciprocal relationships you mentioned and thus losing the possibility of reading their richness, their complex mutual necessity within the framework of the transformations of the built landscape.

The rhetoric of the void that has long been in use among architects (urban void, non-place, disused places, etc.) can be read, in this way, as produced by the misunderstanding of the concreteness of the space-interval between forms, of their fundamental role in establishing necessary, if conflicting, relationships between things, by the non-unified investigation of the forms of the city and the territory that reduces every object to fragment, every form to ruin, that is, construction separated from its organic dwelling. There is no non-place except in the abstracting image that separates space from its own contours, which in-form it, give it form and meaning.

This rhetoric does not actually address the problem of the complexity of urban phenomena. It only makes them complicated by making the architectural project lose vitality. In fact, the limitation and, at the same time, the strength of the architectural project lie in its exemplifying and simplifying function of reality. Function that constitutes, it should be noted, anything but the trivialization of the data of the built world. On the contrary, simplification represents the critical act of recognizing the backbone of problems by allowing us to overcome the accessory, the detail, the secondary complement destined, in the project, to conform (to compose itself in unity) according to the general trace drawn in the insidious tangle of signs that every landscape present. This unity, inherent in the notion of organism, is the tool, moreover, that allows us to dialectically understand the fragmentation and laceration of the contemporary world.

R.S. How can your ideas of organism and unity of architecture, which give the title to one of your books on design theory, be explained in the current condition of the metropolis, usually interpreted as an assemblage of conflicting parts. Moreover, the most widespread architectural literature interprets the condition of the present space as having an immaterial life, a kind of second life, parallel to the physical one, which seems to make the material relationship between things quite unnecessary.

This is partly true and needs to be considered, but there is also a lot of rhetoric in this widespread vision of the future. Pessimistic rhetoric, in my view.

Of course, by definition, the outcome of this conflict, the fragment, is disorganized, it is the outcome of the rupture of an original unity, to which new forms of communication and representation seem to contribute. But in fact, architecture possesses its own non-eliminable physicality capable of accommodating the mutations that affect the environment.

Much more than new forms of immaterial communication, changes induced by mechanized forms of transportation have long contributed to the updating of buildings, fabrics, and structures of the territory. Consider, compared with the relatively little impact produced using the telegraph or telephone, the introduction of new physical vertical mechanical connections and the revolution wrought by the elevator in the form of building organisms. The specialized building traditionally took the form of a fabric in which paths had been turned inside out (the palace, for example, the convent and the endless building types derived from them). They could only be primarily horizontal, as staircases could only pose as communication between "routes" due to the brevity of their development.

With the new physical forms of transportation, the elevator becomes a real axis that identifies a vertical route on which secondary horizontal routes are implanted that distribute the different floors. The logic of the hierarchization of the routes of a fabric is thus reconstructed, vertically, physically opening a third dimension to the increase of the aggregative organism. To consider the high-rising building as a vertical fabric, to study it with the tools of processual morphology and not only to collect, as has been done, its suggestions, would open new frontiers to contemporary design.

Consider then, moving to the scale of the urban and territorial organism, the physical innovation constituted by the introduction of subways, rail networks, and even today's total mechanization. From one place on earth to another, travel takes place today only through the mediation of the machine (the metropolitan car or train, the conveyor belt, the escalator, the airplane, and then again, the escalator, the conveyor belt, the metropolitan train, or the car) in a sequence of routes that acquires its own (recognizable and programmable) distributive, spatial, and aesthetic character. Alongside these phenomena traceable within the general framework of the process of mutation of the typical characters of the city and the territory, the other phenomena due to the formation of immaterial communication networks (the decrease in the hierarchy of centers or de-centralization; the instability of the relationship between productive activity and place, or de-localization) intervene as a general indication that cannot be articulated in all its unpredictable complexity. The new spaces will therefore be "generic," capable of accommodating the multiple.

R.S. You mean a simplification of spaces as well? Generic as "in general," common, shared, but also "generating," able to accommodate complexity?

The design/reality dyad has been traced back to the pair of opposing and complementary terms simplification/complexity, dialectically giving rise to the two notions of simplified reality and designed complexity where the former indicates the elementary knowledge of the function not only of the design, but also of the reading from which the design derives, while the second is associated with the abstracting role of contemporary design, which tends to propose as the end of architecture not the complexity resolution (or at least the rational control) of complexity understood as a multiform reality (as a multiple form in which to recognize the higher degree of unity) but the evocation of fragmented unity as an inextricable plexus. In this sense, the theorizing of complexity researchers such as Rem Koolhaas (and the dense host of epigones who in the wake of their indications, in an overt form, contributed significantly, in the late 1970s, to the contemporary interpretation of the role of the architect) has been posited as a simplified reading of reality that is followed by an artificial design of complexity.

Koolhaas's reading of New York was avowedly "a retroactive manifesto of the construction of Manhattan, the Rosetta Stone of the XX century": all the economic and social contradictions that led to the formation of a gigantic inextricable territorial knot, terrible and fascinating, were succinctly grasped in their aesthetic value of identifying urban complexity, to be reproposed, with the tools of description, and therefore literary, in the design of the new, even in cultural areas with radically different characters.

Manhattan was read as a place deputed to permanent transformation that contains not only the image of mutation, irrationality and utopia, but also the layers (layers!)

of possible architectures, aborted projects, popular fantasies that constitute possible alternatives to the built reality. Koolhaas's city, a critical interpretation of an explicitly mythologized "congestion culture," seemed to grow in a programmatically disorganized and serial way, by continuous additions struggling for survival. But the evocation of complexity in his imitated urban projects seemed like a thaumaturgic gesture, likened to the shamanic ritual that mediates mysterious and uncontrollable forces.

In fact, the loss of contact with reality, the process of increasing abstraction of forms, also has deep reasons in the new economic conditions that have gradually characterized Western societies since the 1950s. Conditions of extreme affluence. The liberation from the constraints imposed by need, which clearly highlighted the relationships of elementary necessity between things, ended up producing the decay of the nexuses that help explain why and how the parts of a built organism (of a building, but also of a city or a territory) conform, how they develop relationships of congruence between the component elements.

In the context of a general crisis that leads to substituting the quantity of goods produced for the quality of the man-made environment, the rational and aesthetic principle of appropriate proportion of means to ends to be achieved has gradually lost its founding role. The loss of this proportion has also meant the loss of the ethic of the good use of resources, which coincides, in large part, with the art of knowing how to build well. Now we are coming to terms with the consequences of this situation, with this waste of resources that has produced considerable failures, of which architecture is the timely representation.

R.S. In this dramatic picture of the contemporary condition of design and the role of the architect, what do you think might be a possible way forward? If architecture is necessarily an expression of the society that produces it, are we doomed to have projects that waste resources for the sake of media spectacle?

The architect should also have a critical role with respect to the society in which he or she operates, but this is not happening.

On the other hand, claims for change toward the wise use of resources, toward their organic coordination with human life, are becoming more and more pressing in recent years, especially from the younger generation. The organicity of the built world, the search for congruence between its parts, is turning from requirements imposed by necessity in past centuries into a conscious choice. Perhaps, they are even becoming a new value in architecture, a discipline eternally lagging the times.

In the general confusion of things, we are beginning to understand the fundamental role played by the historically unprecedented boundless squandering of wealth in some societies of the Western World due to an availability of resources thought, for some time, to be inexhaustible.

From this organically optimistic view of the world, we could try to imagine a new architecture.

It is possible that, in the face of new environmental crises due to wastefulness, past the naive euphoria and amazement at the changes taking place, the architecture of the future city will be composed of large empty compartments, where sometimes the footsteps of a single solitary visitor will resound, and at other times the din of the great flows of an unstable, noisy, migratory population will burst in. But the architect will not conform the rooms to the silence of one or the sonorous excess of the others. Anything that is destined to adapt to a mutation can only disregard the forms of mutation, having to possess a higher degree of solidity, capable of accommodating the different and the unknown.

The shipwreck of the myths of the "human dimension," of the many machines for inhabiting, caring, gathering, has shown, beyond any doubt, the need to separate the mutable, the accessory, the transient (the mobile and fragile part of the built) from the solid durability of the plastic structures that support and separate. This architecture will give its exemplary products precisely in large infrastructures in contact with speed and the machine. In factories where ever-changing technologies do not allow form to follow unstable, ever-changing functions, but also in large nodes of territorial exchange. In all those places, in other words, where complexity comes from a lack of recognition of the connections between things. Recognizing the airport or the subway station as nodes in an architecture of routes is indeed difficult in the face of the lack of physical visibility of the network itself. And yet it would be simplistic to read the airport as a fragment, the subways stations as scattered tiles in the flow of urban transformations. On the contrary, architecture must also take on the non-visible, the not directly legible organicity that binds spatial nodes together. The notions of organism and congruence will re-emerge precisely in the contexts from which, as recognized expressions of modernity, they would like to be furthest removed.

N.S. Focusing on the idea of "knotting," one of the notion through which the contemporary idea of organicity is better expressed, I would like to ask you how, in some of your projects, as for example the Quebec National Assembly building, this notion has been declined at the scale of the architectural organism?

G.S. This project was done in 2016 and was conducted within the *École d'Architecture* of the University of Laval. The building in question is an expression of the cultural and political identity of Canada's vast enclave of French culture. Built at the end of the XIX century, organically in direct contact with the building fabric of the city, it was progressively isolated by the construction of new complexes for institutions which destroyed the existing urban structure. Indeed, it came to form a "city of politics" detached from the physical context in which it is located.

The building develops around an open space which in fact no longer forms a courtyard, but rather a sort of partially infilled waste area (Figs. 4.2 and 4.3).

The project proposes the transformation of the central open space into a new, large covered urban node (Fig. 4.4), in which routes coming from the outside intertwine to form a place dedicated to the gathering of the citizens, addressing the theme of the relationship between public institutions and urban space.

The proposal addresses the relationship between public institutions and cities, which runs through the entire formation course of major national capitals and presents today, not only in Europe, particular urgencies due to the difficult relationship developed between citizens and politics (Fig. 4.5).

Project E: The city of politics. Knotting of the National Assembly Palace in Québec City. 2017

G. Strappa (coordinator), P. Carlotti, with V. Buongiorno, G. Emmi, C. Sammarco (PhD students).

Politics as central theme. *The ideas developed during an international design workshop in Quebec City, revolves around the correlation between state institutions and city fabric. They present very peculiar urgencies today due to the difficult relationship between citizens and politics, but also to the symbolic role that these places often have and to new functions they are assigned, destined not only for the discussion of ideas but also for their communication and diffusion.*

As part of the collaboration between two study groups (Italian and Canadian), the aim of this proposal is to find a solution to the unresolved urban node of this building (Figs. 4.2 and 4.3).

Design proposal. *The project is based on reading and interpreting the forming process of the urban fabric, ending in a specialized, functional/esthetic synthesis of it.*

The relation between space and limit takes form by transitioning from an open space into a covered one. The design idea is based on the notion of "knotting" (Fig. 4.4), *an ideational process that underlies the formation of many building types characterized by the presence of a large central nodal room served by serial rooms on the perimeter. In this case, the large room is conceived as a new space available to the community, reached by urban routes that intersect within it, penetrating the body of the enclosure creating a new internal and collective spatial quality* (Fig. 4.5).

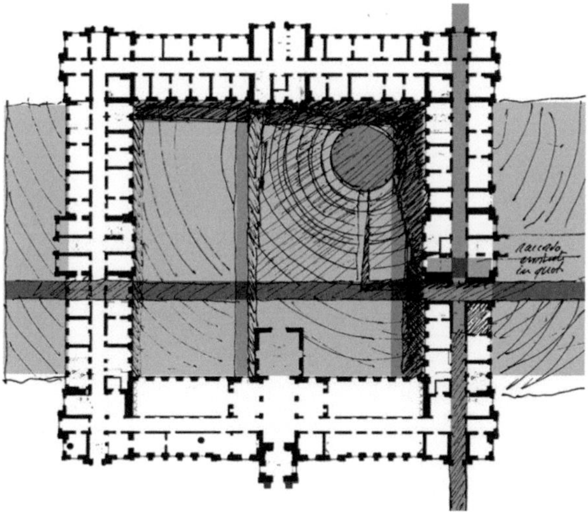

Fig. 4.2 Ground floor sketch of the of the proposal

Fig. 4.3 Sketch of the inner courtyard transformed in a nodal, covered space, 2017

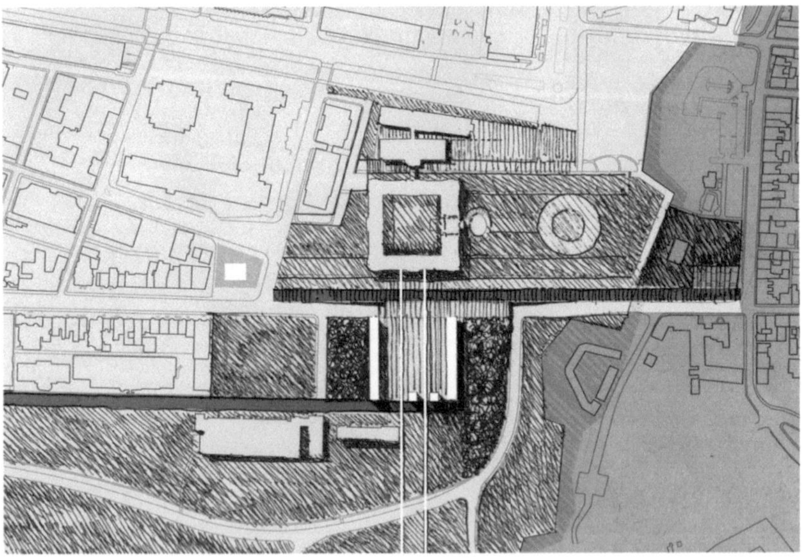

Fig. 4.4 Transformation of the Quebec National Assembly Palace as an urban knotting

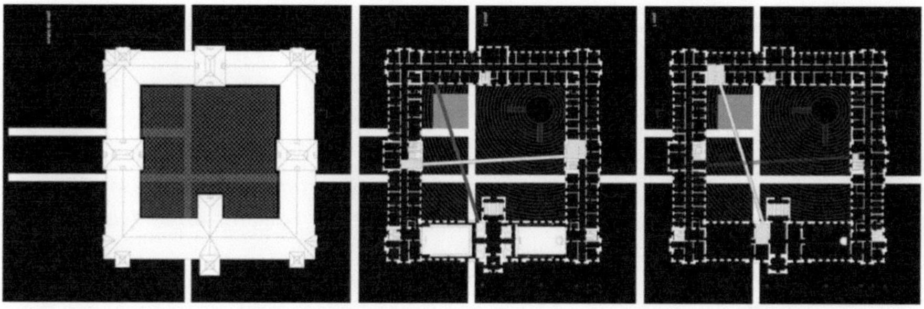

Fig. 4.5 Roof, ground, and first floor plans

Forming process of the new Quebec National Assembly building

First phase: first construction intervention, in 1877, of the National Assembly Building around a central courtyard
The building is progressively isolated by the Honoré-Mecier and Grande Allée street. Its relation with the urban fabric is gradually lost.

Second phase: progressive isolation
Construction of two new buildings to the north east of the Palace, along Rue Joly del Lotbinière which contribute to separating the building from the urban environment. A complementary geometrical garden is formed.

Third phase: existing state.
The housing fabric is replaced by large-scale special buildings, such as the Hilton and Marriott hotels and particularly the Secretariat building, completing the process of isolating the National Assembly building

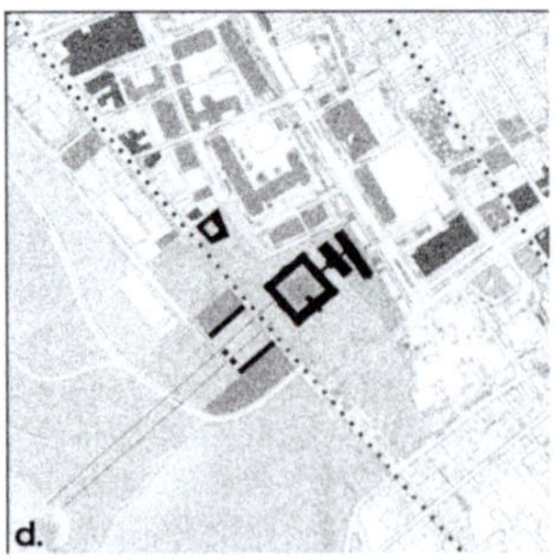

Fourth phase: transformation hypothesis.
"Knotting" of the Palace with the freeing of the courtyard and coverage of the open space (Fig. 4.6).
Demolition of the Secretariat offices, replacing it with two linear buildings ("lames") that frame the view towards the river and create a large public space for events and public activities.
The part of the Grande Allée in front of the building is buried to connect it directly to the Plains of Abraham.
Implementing green spaces

Fig. 4.6 Forming process of the building fabric and the new design as last transformation phase

Bibliographical Note

Berman, M. (1982). *All that is solid melts into air. The experience of modernity,* Simon and Shuster, London

Caniggia, G. (1989). *Permanenze e mutazioni nel tipo edilizio e nei tessuti di Roma (1880–1930).* In G. Strappa (ed.) *Tradizione e innovazione nell'architettura di Roma capitale. 1870–1930,* Kappa, Rome.

Caniggia, G. (1997). *Valori e modalità del restauro: valore storico e valore architettonico. Relatività e consumo dell'opposizione dei termini "vero" e "falso",* In G. L. Maffei (Ed.) Gianfranco Caniggia, Ragionamenti di tipologia, Alinea, Florence.

Descartes, R. (1637). *Discours de la méthode pour bien conduire sa raison et chercher la vérité dans les sciences,* Jan Maire, Leyde.

Giovannoni, G. (1903). *I restauri dei monumenti e il recente congresso storico.* In Bollettino della Società degli ingegneri e degli architetti italiani, Vol. XI, 19.

Giovannoni, G. (1929). *Questioni di architettura.* Biblioteca d'Arte Editrice, Rome.

Giovannoni, G. (1931). *Vecchie città ed edilizia nuova,* Unione Tipografica Torinese, Turin.

Hitchcock, H. R., Johnson, Ph. (1932). *The international style: Architecture since 1922.* W.W. Norton, New York.

Ieva, M. (2021). *Morfologia urbana e linguaggio nell'opera di Gianfranco Caniggia,* FrancoAngeli, Milan.

Muratori, S. (1980). *Storia e critica dell'architettura contemporanea. Disegno storico degli sviluppi architettonici attuali,* Centro Studi di Storia Urbanistica, Rome.

Strappa, G. (1995). *Unità dell'organismo architettonico. Note sulla formazione e trasformazione dei caratteri degli edifici,* Dedalo, Bari.

Strappa, G. (1998). The notion of enclosure in the formation of Special Building Type. In Petruccioli, G. (Editor) *Typological process and design theory,* Cambridge (MA)

Strappa, G., Ieva, M., Dimatteo, M. A. (2003). *La città come organismo. Lettura di Trani alle diverse Scale,* Adda, Bari.

Strappa, G. (2003). L'eredità progettuale di Gianfranco Caniggia. In C. D'Amato, G. Strappa (Editors), *G. Caniggia, dalla lettura di Como all'interpretazione tipologica della città,* Proceedings of the International Conference held in Cernobbio on 5th July *2002,* Adda, Bari.

Strappa, G. (2003). Gianfranco Caniggia's notion of organism and the heritage of the Roman School of Architecture. In G. L. Maffei (Editor), *Gianfranco Caniggia architetto,* Alinea, Florence.

Strappa, G. (2006). *Lettura e progetto dell'organismo urbano di La Valletta.* Polibapress, Bari.

Strappa, G. (Ed.). (2012). *Studi sulla periferia est di Roma.* FrancoAngeli, Milan.

Strappa, G. (2015). *L'Architettura come Processo. Il mondo plastico murario in Divenire,* FrancoAngeli, Milan.

Chapter 5
Territory

Abstract In this chapter, we talk about the territory as an architecture in its fullest meaning, arising from the collaboration between man and nature. Also Territory is *materia signata* (marked matter) a substance to which man's consciousness recognizes an aptitude for transformation. The territory is considered as a "territorial organism" and its processual meaning cannot be understood unless its elements (routes, settlement, productive areas) are placed within a frame of relationships of necessity between them. Landscape is defined as the visible aspect of the structure of a man-made territory. Understanding the structure of a territory through the phases of its anthropic use should begin with the reading of the routes, the way in which they form, consolidate, specialize, and hierarchize among themselves in a cooperative way, according to relations of congruence and proportion with the settlements and productive areas to which they fit in. As for contemporary conditions, the land is considered today the place in which architecture is "situated". Territory, itself an architecture, and it is necessary to consider its actual disaster as a state produced by crises, conditions of rapid, traumatic change, even climatic. And also produced, by the fact that we no longer know how to read the territory in its architectural structure. Differently from the general meaning of the term, it is proposed to give the word "Anthropocene" a cultural sense: "Anthropocene is the epoch in which man began to completely transform the earth without reading its form."

Keywords Materia signata · Territorial organism · Anthropocene

R.S. Territory is perhaps the most complex topic among those with which architects are concerned, not only because of the scale and the large number of components that form it, but also because its problems seem to overlap, more than in other fields of design, **with issues of economic planning, political management, and social organization.**

Inhabiting a territory seems, therefore, a far more complex issue than inhabiting a house. Is it thus true that at this scale, which includes all others, it is still possible for architects to start with the study of form?

R. Salamouni et al., *Architecture in the Making*,
https://doi.org/10.1007/978-3-031-86833-7_5

G.S. Sure, I believe we have, as architects, to read and design the territory through its form. That is our job, our specificity. Other disciplines study the territory with their own methods and according to their own statutes, but we have to study it through our own procedures, even if collaborating with other fields of knowledge.

Territory is, in fact, architecture in its fullest meaning. Arising from the collaboration between man and nature, it is therefore made up of the soil plus the work of man. As for any architecture, the beginning is always recognizing: distinguishing the qualities and availability of the soil, identifying its inclination to be transformed according to certain purposes.

Soil is the matter of the territory. Just as man recognizes in natural matter (the rock, the clay, the tree branch) the possibility of being transformed into material (blocks, bricks, beams). So, the man recognizes in the different composition and shape of the soil the opportunity of obtaining paths, routes, courses, productive areas and places suitable for settlements (Figs. 5.18 and 5.19).

Thus, land too is *materia signata* (marked matter) a substance to which man's consciousness recognizes an aptitude for transformation: matter aimed at the construction of a space suitable for life, at its "settlement," a term that well expresses the meaning of the processes of formation of the built environment. "Settlement" means the act of to settle (from Old English setl, a place to sit), that is, to stop and take possession of an area to make it suitable to inhabit, another important word related to possession. To inhabit derives, in turn, from the Latin *habere*, to possess with continuity (Figs. 5.20 and 5.21). So, the term expresses the intense connection between inhabitant and inhabited space is: a space in which one not only stays, but which one knows deeply, where one lives and often, together, works, as in the row-house widespread in many European territories and cities.

Seen in this way, the house is a small world, placed in a somewhat larger world that is the *contrada* (the elementary aggregative organism) which, in turn, is placed in the complex sphere of the urban organism and, finally, in the territorial universe. Therefore, the house is the origin of the built space and the whole built world derives from it by aggregation/specialization.

A settlement is then a temporary or stable structure consisting of a set of dwellings organically related to a complementary productive area. The first temporary settlements have traditionally been considered as belonging to the civilizations of hunters and gatherers, followed by semi-stationary settlements, related to the earliest forms of cultivation or breeding, where the notion of dwelling (from Middle English *dwellan*, remain in a place, with the sense of permanence) is more firmly associated with that of pertinent area. In fact, the stable possession of an area is originally due to the appropriation produced by the work that is done there with greater or less continuity. It is to be assumed, in this regard, that the first, embryonic stable forms of pertinent area can be associated with the anthropic transformations of the Neolithic age, with the spread of cereal crops, due to the need to overcome the instability of simple sporadic harvesting.

But the ultimate association of a natural soil with man-made structures built to use it, (that is with stable anthropogenic modifications), is to be found in those historical phases and areas where the need for continuous maintenance of the soil entailed a greater custom between man's work and the productive area, as in the territories where the conditions for cultivation had to be produced artificially through permanent irrigation systems involving as much organic collaboration between man-made works and natural soil as they did organizational collaboration and specialization between work groups within the community.

This act of domesticating natural soil conditions, concerning transformations to be made on nature not simply "encountered," but "chosen," is a typically architectural operation: man "reads" the shape of the soil, recognizes its characters, makes a selection according to the aptitudes of the different parts (the potentialities of the ridges, the unknowns of the valley floors, the possibilities offered by the waterways) and then specializes them (Figs.4.1 and 4.2). As in any construction, at any scale.

But the territory is also a great civil legacy, a heritage in which are inscribed the choices and culture produced by the people who have lived there. That is why any intervention at this scale obliges an understanding of the characters we can know through its form, knowing that this form is the recognizable outcome of an ongoing process.

Since the territory is made up of cooperating parts, it is possible to read it through the notion of organism, that is, a set of elements, structures, systems linked by a (time-varying) relationship of necessity. And as in every organism, it is possible to recognize in it historical phases and cycles that determine its formation, transformation, fragmentation, ruin.

We grasp this process through momentary states of equilibrium that restore a discrete idea of a historical sequence that is, instead, a continuous flow of modifications and updating.

This is why the historical-processual meaning of an urban organism, or a system of routes cannot be understood unless their form is placed within a relationship of necessity with the set of interactions established in time and space within its territorial surroundings.

This form of the man-made territory, the visible aspect of its structure of relationships that binds in the notion of organism the different scalar degrees of the built environment, can be referred to by the term "landscape."

R.S. If our purpose is to study the territory through its form, its reading and representation become central. You wrote that any graphic description of the territory cannot be neutral as it contains a critical substance due, if nothing else, to the fact that information deemed significantly is represented at the expense of other secondary information. What then is it about cartography that we as architects are interested in?

G.S. While of a building we can have a synthetic representation through sight, of the territory we cannot have an equally general and summarizing physical perception. The earliest images of territories are ideal configurations, ordered by an intellectual or spiritual desire for form. They are unitary representations of a cosmos, a Greek word (κόσμος) meaning "order." Figures destined to remain abstract and symbolic for a long time.

It is no coincidence that the irruption into the Western imagination of its physical, concrete image, began to occur in the XVIII century thanks also to new techniques of observation. The inventions of the hot air balloon, the submarine, the exploration of mountains and deserts contributed in generating new artistic categories related to the idea of the sublime (the abyss, infinite space, etc.).

But the technical representation of the territory continued to be performed symbolically. Cartography was, and is, the instrument of this representation, constituting the result of critical reading, the reduction of notions deduced from observation, communicated through symbolic forms.

The symbol (from Indo-European *symballein*, to unite, to put together) expresses in a condensed way the notions considered fundamental: it is thus itself the result of choice and selection. For this reason, the cartographic tool (the symbols and the structure that binds the symbols together) is in direct relation to the author's interpretation, within a cultural area and a particular historical phase, of the territory he represented.

Just as it is possible to trace the lines of a process of transformation of the territory, so it is possible to trace the lines of a process of transformation of cartographic tools. They testify not only (and not so much) to the changes in the characters of the territory itself, but, above all, to the knowledge and consciousness that, in different historical phases and in different civil contexts, man has had of his built environment.

For example, the ancient Tabula Peutingeriana, which has come down to us in a copy from the XII–XVIII century, mainly shows the structure of routes in the Roman Empire; in Giovanni Magini's atlas of Italy drawn in 1620, on the contrary, land routes hardly appear; in a modern automobile map, the road network takes over the representation of other observable data on the territory; finally, the plan of a metro is mainly the symbolic and abstract drawing of a network, with poor relations to the physical form of the routes, to the actual dimensions of the distances between the connected poles.

For those who, like the architect, need to read the territory in an active way, having as an end the intervention, it is therefore fundamental to extract and interpret the available information with coherence, understanding that the available representations must be interpreted.

It is necessary for us, therefore, to establish a reading criterion related to the notion of territory understood as an organism, through the synthetic exposition of its formation, the representation of its structures, and the operational tools that allow us to extract from the different types of representations the notions considered useful.

The interpretation of the structure of a territory, identified through the phases of its anthropic use, can only begin with the reading of the routes: from the way in which they form, consolidate, articulate, specialize and hierarchize among themselves in a collaborative way, in a relationship of mutual necessity, according to relations of congruence and proportion with the settlements to which they fit in. The motion of man on the ground, the marks left by displacements, crossings, migrations, precede, in fact, any other trace: any structure linked to work activity and appropriation is associated with routes, more or less stable, that originate it and allow its existence.

Routes, like any datum of built reality, possess their own typicality, that is, they are recognizable, in their formation and evolution, through common characters that historically identify their phase of belonging and, physically, their pertinence to the natural characters of the soil with which they are associated.

Original routes are, we could say, the most ancient interpretation of the territory: human beings traveling great distances, read in ridges the opportunity of traversing vast areas without the obstacle of rivers and swamps that occupied the valley bottoms, then organizing secondary ridge routes that allow reaching habitable and defensible areas at the end of promontories overlooking plains rich in potential resources.

Of course, the territory has been much transformed since the industrial revolution. But it is also true that in recent times a veritable literary genre has developed about its disintegration. Its captivating aesthetic significance ended up organizing consensus on the current condition. I believe that the reading of formative processes can show that even the territories of contemporary expansions of large urban centers, often referred to as "informal," possess a form linked to their own history. They are historic territories too, that should be read as the promise of a new organicity.

R.S. Most of your projects seem to be located in western contexts, in Europe more specifically. These countries share comparable territorial features. However, eastern countries group a large scale of different landscapes. In your opinion, the same processual methodology for reading the territory could be used in other cultural areas? How does it impact its architecture?

G.S. Even for the territory, the principle on which to base the morphological study is to look for the general in the particular. This means that there is a part in the methods we use which is universal. In some ways man, his needs, his way of standing in front of nature and transforming it according to his needs are quite analogous with exceptions, certainly, all over the world.

The elementary actions that underlie the construction of historical territory (displacement, settlement, work) have many similarities in Europe, in China, in South America. Historic territories are not only those with long traditions, as often in European countries, but also the large conurbations on the periphery of the world. The example of the formation process of many favelas, whose form interprets the unraveling of ridges and valley floors, possessing some spontaneous rules that liken them to the formation of some medieval European settlements, is valid for all.

But it is also true that this "general" that the subject seeks also depends on the character of the object of the search, the particular that is given in different forms, and then it also depends on the subject's culture, his experience of the built world, his idea of process (of historical time, of future time).

The theories we have developed to interpret landform are based on soils formed by continuous ridges from which branch off, in different forms, secondary ridges that run along habitable and inhabited promontories. There are thus directions in the land-structuring process: from the high altitudes, where the most continuous ridges exist, but lacking in springs, toward the valley, which is more difficult to domesticate, but rich in water and potential nodes of exchange, of future markets. The Apennine ridge and its branches to the Tyrrhenian and Adriatic seas are an example of this theory, but also (see the Muratorian studies) the origin of it.

These criteria of investigation have been successfully applied in other parts of the world such as, for example, Algeria, where the orography has favored the development of processes having remarkable affinities with the Italian one. But also, significant geographical differences, due to contact with inland desert areas, and then historical differences (think of the trauma of European colonization and the formation of non-processual structures), technical differences, such as with exceptions, certainly. Here to speak of ridges traversed to form the first settlements is problematic because, as Alessandro Giannini has written, it is necessary, more than to look for fertile half-coasts, to see the great river as "the only duteous connection" between Central Africa and the Mediterranean, traversed by slow but continuous currents of people. A "ridge, in short," to quote him.

There are also macroscopic differences in the object of research. The most obvious case is the formation of the territorial structure in the islands and in all areas where territorial configuration was formed from landings due to migrations or invasions of areas still without established structures, as in the evident case of Malta. Then there is the specificity of the reader's culture. We tend to see realities other than our own with our own parameters and culture, but this is not necessarily always an awful thing. It is our truth, which is useful if we admit that there may be several others.

Saverio Muratori's notion of territory originated in interwar Italian culture, with its specificities. The idea of fringe belt, originated from the hypotheses formulated by German geographer Herbert Louis in the 1930s and developed by his student M.R.G. Conzen in the 1960s, applied to the study of the Chinese city, has been an interesting tool for interpreting, for example, walled cities as demonstrated by the exemplary case of the city of Nanjing.

R.S. Could you go through your views on the notion of Anthropocene? The definitions that geologists normally give of it concern problems that seem to be beyond the architect's control. Is it possible to give an "architectural" definition of it?

The one posed by the notion of the Anthropocene is, for us, a broad problem concerning the role the architect has played throughout history, and the role he plays now, in the context of the society in which he lives. Leon Battista Alberti said that the center of the architect's profession is not to design or to build, but it is to choose. All the architecture of the Modern Movement has, in fact, been oriented by choices, often radical ones. What is significant in the work of Le Corbusier or Wright is not only the extraordinary formal innovation of their architecture: more important is the critique of the contemporary world from which derives a progressive vision of society, a "passionate" desire for transformation and thus for a form that expresses it.

Form is a solution and a conclusion.

Today, on the contrary, there is no critique, no real choices. Architecture conforms to economic and political power, provides what the market demands to the architect. The spectacular, the unprecedented: revolutions without center or explanation. At all scales.

It happens with the territory also. Architects witness the of the territory, evident to all, as an inevitable natural phenomenon, a disaster in the etymological sense: disaster "due to the bad influence of the stars."

Today, the condition of the land is considered the given of the problem and architecture is placed in it.

It is not. Territory is itself, as we said, architecture, and it is necessary to consider disaster not as something that depends on the stars, on fate, but as crises, conditions of rapid, traumatic change, which induces thinking about the future.

Unlike Paul Krutzen (who popularized the term and was an engineer devoted to atmospheric chemistry) I would give the Anthropocene a cultural rather than a scientific meaning: in my opinion, "Anthropocene is the epoch in which man began to completely transform the earth without reading its form."

The center of the problem, as you see, is still the form. I believe, from this point of view, the proposals on the date of the beginning of the Anthropocene are interesting. As is well known, the proposals are many and date from at least the XIX century, when Antonio Stoppani, in 1873, began to define man as a "telluric force," calling ours the "Anthropozoic era."

For the Italian situation the proposal to consider 1945 as the beginning of the Anthropocene is particularly interesting, not only because the first atomic bomb was detonated, but because in that year World War II ended in Italy. The period of Reconstruction began in those years, with the rapid transformation of the territory. Large infrastructures and highway networks were realized shortly afterward. These were the years of mass motorization and the baby boom.

The modernization of still pre-modern regions seemed to begin. It was a great opportunity to give concrete form to a coherent idea of territory. Instead, a seemingly pragmatic way was pursued, following only demographic and economic development without any attention to the shape of the soil, to its forming process.

The lack of a clear legislative framework together with a bureaucratic territorial governance lacking general perspectives were the preconditions for the "natural" calamities that occurred within a short time of the start of the Reconstruction. At that time, new construction was taking place everywhere, often in historically at-risk areas. Rampant deforestation was causing dangerous conditions. Riverbanks were being channelized and covered with concrete slabs, no care was taken for the hydrographic system, which was endangered by the huge extraction from riverbeds of aggregates that fed the construction explosion of large urban areas.

On October 25, 1954, the first major disaster occurred, when one of the most serious floods in Italy's recent history struck the Campania area of Salerno and the Amalfi Coast. Many other disasters followed, up to the present days. Exemplary is the 1998 Sarno and Quindici landslide. The area of this region consists of a base substratum composed of limestone rocks, above which the various historical eruptions of Vesuvius have deposited successive layers of pyroclasts: two sets at different densities, held together by vegetation and terrace cultivation practices. The abandonment of them meant that the same continuity between limestone and pyroclasts was not any longer assured.

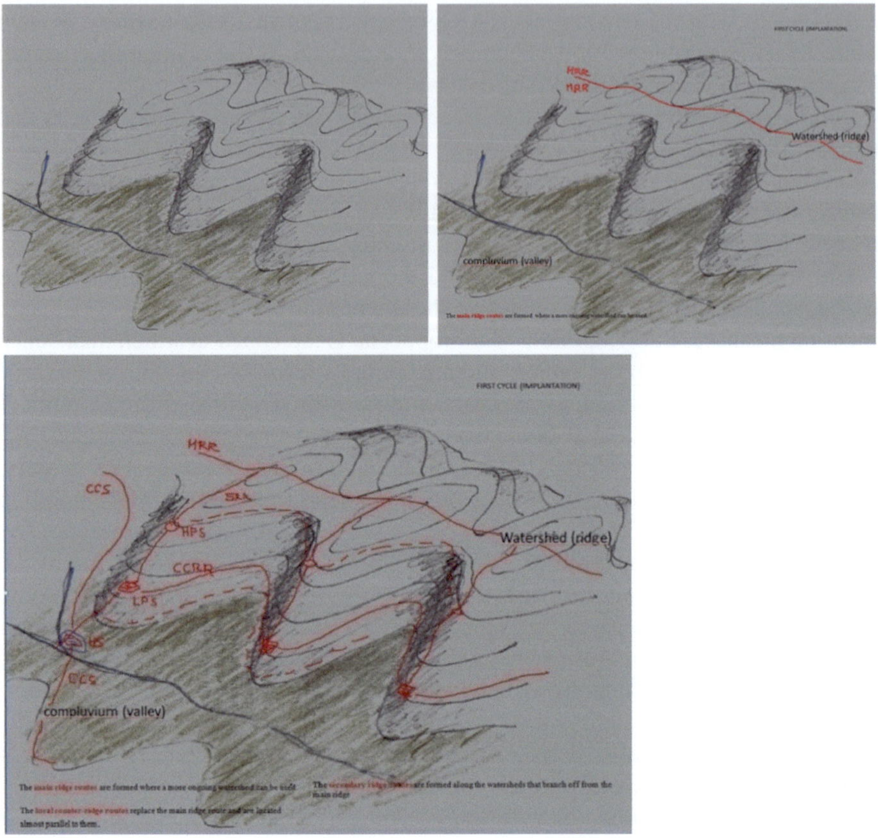

Fig. 5.1 Pictures from the Urban Morphology course taught at the Faculty of Architecture, Roma Tre University, in A.A. 2023/24 by Prof. Giuseppe Strappa: sketches of different phases in the first territory formation starting from a main ridge route (above pictures); examples of low promontory settlement formation starting from secondary ridge routes in the Latium region (below picture)

The old photos of that period show how they had built against any formative process, not considering the shape of the ridges, slopes, valley floors. Disasters all due, in different ways, to misunderstanding the shape of the land and its possibilities for use. That is why I insist that, even at this scale, it is necessary to study the formative processes, the solidarity between landform and anthropic transformations that, alone, can be the premise for land care (Figs. 5.1 and 5.2).

N.S. I'd like to continue our conversation by talking about the notion of scale. The proposal for "The Expo after the Expo" seems to be a special case in your work as a designer, dealing with the "unsure dimension" of a demonstrative experiment in the urban fringe belt.

G.S. The project "After Expo" was developed within the Sapienza Draco doctorate, back when I was directing it (Fig. 5.9). It is quite different from other projects we talked about. It is an experimental proposal, inserted in the Milanese

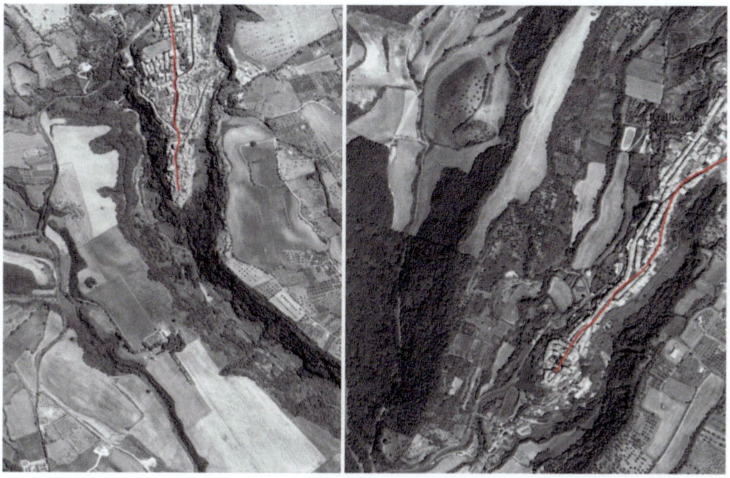

Fig. 5.2 Pictures from the Urban Morphology course taught at the Faculty of Architecture, Roma Tre University, in A.A.2023/24 by Giuseppe Strappa. Low promontory settlements in the Lazio region located at the end of secondary ridges. The territorial route becomes a matrix route at the urban scale, organizing the building fabrics inside the walled settlement

hinterland and an extreme example of the non-deterministic role of morphological reading. In fact, by addressing the issue of large dimensions, the proposal accepts the need for an evident discontinuity within the general condition of crisis in the Milanese suburbs, recognizing that not everything we have historically inherited has a positive value (Fig. 5.10).

The project was supposed to contribute to solving the problem on how to reuse the area on which the pavilions of the 2015 Universal Exposition of Milan were built. An operation with great media success, but a clear urban strategy mistake. The area was isolated amid a node of territorial routes with which it had no relationship. Contrary to common sense, the plan did not envisage the use of the exhibition structures once the event was over (Fig. 5.11). Seen as a redevelopment of a large area segregated from its own territory, the problem was insoluble. Bridges and new roads would not have been enough to transform it into a new element of the territorial organism.

We soon realized how geographical and historical reasons explained the special condition of this place. The portion of territory now included between Bollate, Rho and Settimo Milanese, is crossed by ravines in which flow streams sometimes now lost inside the industrial sprawl and the confused housing expansion. The presence of pseudo ridges partly explained the shape of the routes.

The question that the intervention posed, and its possible solution, progressively evident only after the patient reconstruction of the mosaic of the first 1897 cadaster. In fact, it clearly showed how the apparent disorder of the fabric and the overlapping of the routes over time had followed some principles imposed by anthropic needs and by fragmentary planning intentions. Also, they indicated a close solidarity between the consumption of the ancient substratum, new needs that arose with the irruption of modernity and the shape of the soil, formerly organized by the Roman centuriation.

In this general framework, the Expo location formerly constituted the geometric hinge between different centuriated areas extending to the edge of the north-western strip of the Milan hinterland. The XIX-century maps of the pre-unification land registers still identify them as Pantanedo (quagmire), a place of marshes and malaria, unsuitable for life.

It was evident how, in the succession of the formative phases, new territorial routes had overlaid themselves on the Roman layout to directly connect the small urban centers. On this territory processual development, corresponding to the decay of the ancient form, was superimposed a modern network of large territorial infrastructures unrelated to the local settlement logic (Figs. 5.12 and 5.14).

Therefore, to solve the problem, it was necessary to link this world of fragments with a structure that went beyond the limits of the intervention area, organizing the intersections of routes as real new urban knots (Fig. 5.3).

In other words, it was necessary not so much to mend the ruptures as to give this marginal area a new coherence through a leap in scale. The reading indicated how the settlements of Baranzate and Pero had the vocation of possible territorial poles.

The project proposes, demonstratively, to connect them through a "restructuring route" crossing the whole Expo area. Thus, the formation of an axis organized in a continuous "vertical fabric" was proposed, in which to concentrate new functions. This innovative and peremptory sign would allow the problem to be posed, almost like a manifesto, in an immediate and perhaps not reassuring way.

But it is also evident, I suppose, the design intention to metabolize the modern, often abstract tradition of Le Corbusier, Mayekawa, Reidy research, where, however, the lesson of great urban visions is seen in a framework of territorial realism.

The knotting process, a notion with which I have often dealt with in my projects, is the fundamental datum of the proposal: the experience of large housing interventions has, in fact, taught me how linear structures of this type (valid for all the experience of Corviale in Rome), if not polarized by connections with the real built environment, are destined to rapidly decline.

The design synthesis that the reading has generated, in my opinion, a great unitary sign in the territory, therefore required, more than in other occasions, a hiatus with the formative process of the place, but also an innovative form of congruence with the historical process.

N.S. Keeping up on the topic of territory, I would like to focus the conversation on the definitions of the terms territory/landscape.

In this regard, I quote what you said in your paper "Four dyads of Mediterranean architecture": "For architects, territory is a way of looking at the world. It is the reading of the shape of things (of soils, routes, settlements) to grasp their structure, to understand their origins, their possible transformations. Territory is the inseparable whole of soil and man's work that inhabits and transforms it: it is architecture."

The term "landscape" is the recognizable aspect of a territorial structure, its form containing a set of contributions. I think it is essential that you recognize the

Fig. 5.3 Reuse of the Milan Expo 2005 area. Perspective sections on the linear building connecting the new polarities of Baranzate and Pero

existence of the territory as architectural form and, therefore, of a soil intended as the "matter." I am interested, however, in getting into the merits of the relationship between the two terms, territory and landscape.

From your statement emerges a notion of landscape that synthesizes, in fact, a concretion of events, of footprints, signs, memories. In short, it seems to be traceable to the concept of Hegelian "appearance": that is, to a window through which to access to the constitutive dimension of the territory.

G.S. I repeat that we should consider also territory as matter, a substance to which man recognizes a capacity to be transformed and used. Because I believe that this definition is of great educational utility and distinguishes our understanding of the term from that of urban planners. Moreover, it is a definition that is entirely pertinent to our cultural area and explains that of landscape, which I employ in the sense of the form of the man-made territory. It is the visible appearance of a structure of relationships that binds the different scales of the built environment (I believe that, in this, the use of AI, if used well, will help us a lot in the future).

Territory is the product of man's work in its perceivable aspect, whereas landscape (its beauty, its contradictions) is what we read as the result of a process, the synthetic expression of its ongoing construction.

This notion is different from the more widespread notion derived from Anglo-Saxon culture, linked to nature and the tradition of landscape architecture. While "landscape," the shaping of the earth, indicates aesthetic categories dating back to William Gilpin and Edmund Burke, an order based on perception, on intuition from which follow indemonstrable compositional principles with aesthetic and contemplative purposes, the correspondent term "paysage/paesaggio" is associated with the Latin term pagus, village, recognizing, in a synthetic way, a verifiable relationship of solidarity between soil and built human settlement. "Paesaggio" is, therefore, a cultural expression related to the collaboration between the disposition of nature and the artificial transformations that interpret the shape of ridges, valleys, plains and their aptitude to become built environment.

An authentic architectural notion, then, rooted in the Latin world and its humanistic tradition. Every landscape makes sense, in fact, as a "historical landscape," the outcome of a construction process that can be known in a systematic way, by the same method as the reading of every other aspect of the built world. Seen in this way, landscape, like any architecture, possesses its own unique and unrepeatable characters, but shares some typical aspects with others, deriving from an anthropic process that has, inevitably, formative affinities with other processes. In other words, it is possible to read, in the multiplicity of outcomes the sharing of constituent principles (Fig. 5.16).

Landscape is thus not a semblance, if this term is understood in the common sense of outward and personal perception rather its opposite, and a truth. It is interesting the connection you propose with Hegelian aesthetics, which downplays the task of feeling in the work of art by attributing to it the role, precisely, of a sensible manifestation of a reality. From this point of view, the task of the architect-artist who transforms the landscape could be to reveal, through an aesthetic synthesis, the truth of the form of the land.

In the theme of landscape, as you see, one can easily get lost. To answer your question, perhaps the design for the transformation of a large disused railroad area in Mendoza (Fig. 5.15), which testifies to the effort to read the formation of a natural park as an integral part of the urban organism, can give an idea of what I mean by a design intended reading of the historic landscape (Fig. 5.17).

The topic would require much more space to be addressed systematically. I would just say here, briefly, that landscape could also be concerned as a perceptible form of structures shaped by the human action from a natural soil. As such, it is thus construction, inhabited place transformed by successive formative stages, establishing a relationship of pertinence between man and soil: a landscape belongs to a community of inhabitants, but they belong, also, to the landscape they inhabit.

Project F: Updating the Zen Quarter in the Palermo hinterland, 2008

G. Strappa (project leader), Alfonso Giancotti, Luca Reale, Federico De Matteis (coordinators), Francesco Cianfarani, Simone Di Benedetto, Nunziastella Dileo, Michele Filosa, Manuela Pattarini, Giorgios Papaevangeliu, Luca Porqueddu, Emma Prete (PhD students).

The Zen district in Palermo is often considered one of the great failures of Italian experimentation in the field of social housing. Unlike other cases to which it is usually associated, (the Corviale in Rome, Le Vele in Naples) it nevertheless contains in its own founding idea, the assumptions of transformation, in the same way as every planned city that time and needs adapt to the life of the inhabitants.

***The proposed project** considers Zen as a sort of foundation city in which the formative process continues reconstructing the links with the territory and mending the planned fabric.*

The recent modern settlements on the northern fringe of Palermo had, until the '60s, the identity of suburban villages, replaced later by small "satellite cities", unaware of the territorial and urban links which they interrupted. Among these, the Zen quarter plays a peculiar role through a "generative" geometry that evokes the structure of a planned city of ancient foundation. Moreover, the Zen quarter today also shows the interruption of a developing process that the founding act could generate.

The ancient city underwent, in fact, a process of "medievalization," a progressive structural adaptation between fabric and new buildings. This transformation has in the past given rise to the character of the Italian city generating a less regular tissue, based on new paths derived from the consumption of the ancient form (Fig. 5.4).

***A "humanization" process** of the original geometric rule which did not happened in the Zen quarter, isolated within a frame which seems unchangeable. The result was the formation of an autonomous tissue near the Borgo Pallavicini (a 1958 intervention for multifamily and row-houses) independent from the territorial structure.*

The proposal (developed within a PRIN, Project of National Relevance) is a sort of "redesign" of an urban tissue that would be formed along the territorial matrix if the necessary complementary routes would be generated, then continuing the process with the specialization of housing fabric. This new tissue consists of educational services integrated to the cultural ones existing in the quarter, where the space of the courtyards, linked by a route "turned inside" and parallel to the external path, will "knot" forming further specialization to the building scale. The project, part of a more general investigation about the redevelopment of the heritage of public building in the south-central Italy, then traced the lines of a possible new urban design, reading the system of territorial patterns and their possible renewal, as if a new elementary fabric would "pollute" the planned fabric contaminating its rigidity (Figs. 5.5, 5.6, 5.7 and 5.8).

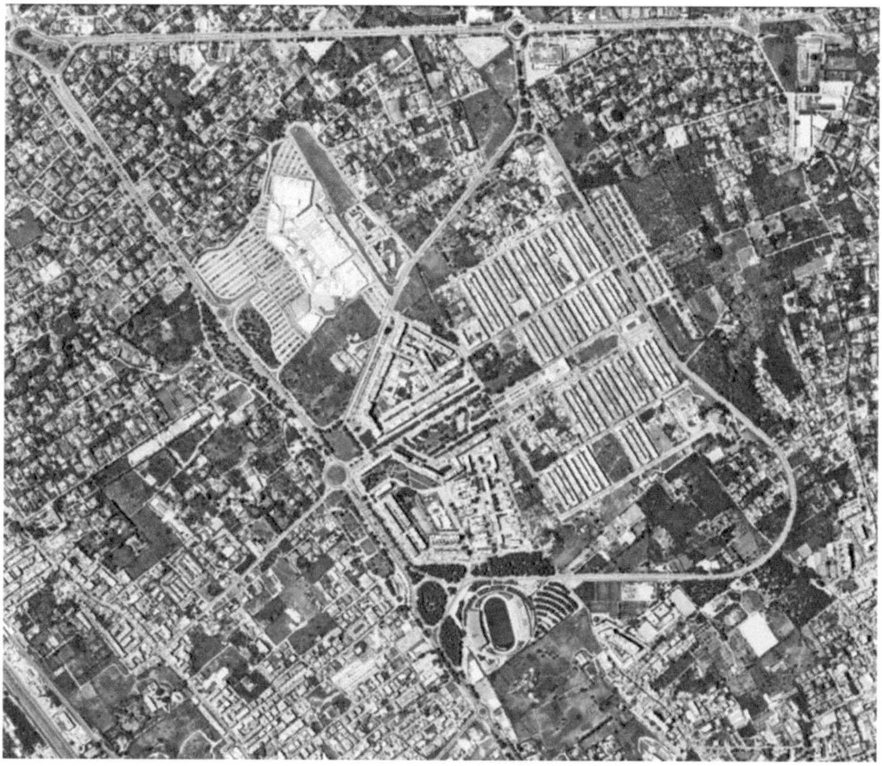

Fig. 5.4 Current situation of the Zen quarter, one of the best known and most discussed public housing projects of the late 1960s. Designed according to innovative principles, it was isolated by the ring road and partly occupied by mafia gangs, constituting a symbol of the degradation of Palermo's suburbs

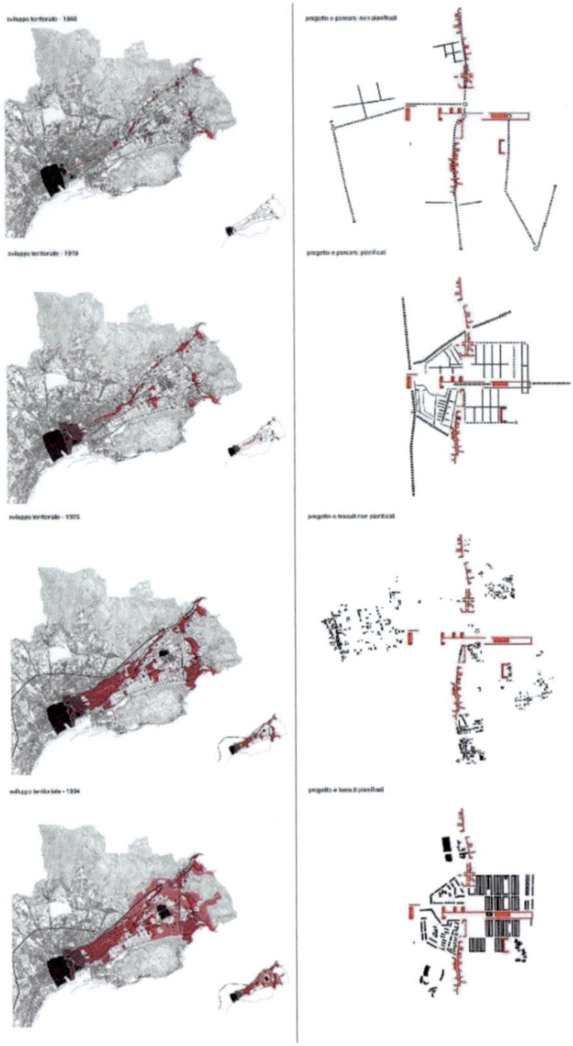

Fig. 5.5 Left. Formative process of the settlement structure in the hinterland northwest of Palermo from 1848 to 1994. Right. Formative process of the project. Overlapping of the different layers: 1. Project and unplanned routes. 2. Overlapping of the planned routes of the 1969 plan. 3. Relationship between service facilities and unplanned routes. 4. Project and planned fabric

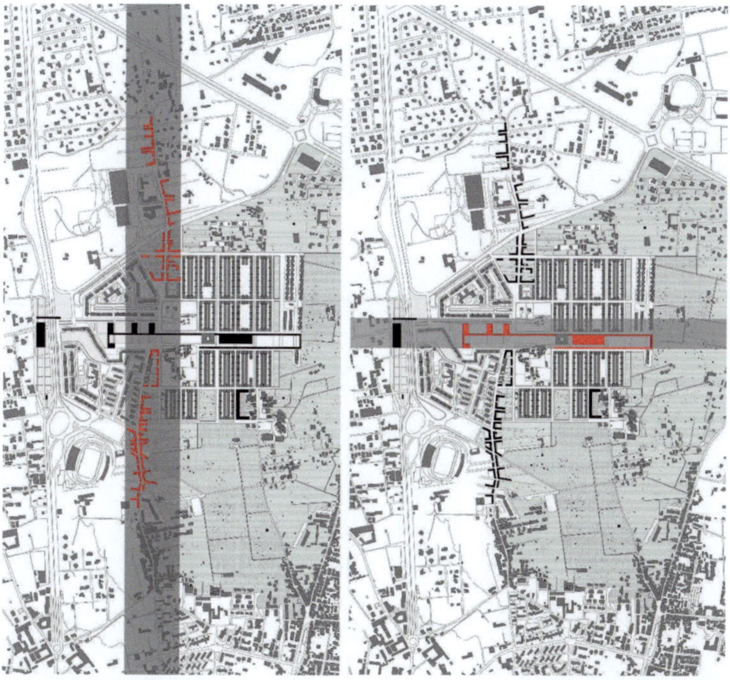

Fig. 5.6 Formation of the urban nodality at the intersections between the planned strip of services and pre-existing territorial route, considered as a matrix route

Fig. 5.7 Architectural design of the didactic and collective services along the matrix route restructured with the introduction of new educational services. The new buildings (middle and high school, library, gymnasium, etc.) are aggregated according to the logic of a fabric, with the serial elements of the various buildings organized along the "building routes" orthogonal to the main (matrix) route

Fig. 5.8 Synthesis of the design proposal with the pre-existing routes and the planned axis highlighted. The outer ring road has been eliminated and the connections between the neighborhood and the rural area restored. One of the secondary territorial routes demonstratively penetrates into the planned grid, generating a nodality of services at the entrance (in red, lower right of the Zen district)

Project G: Expo after Expo. Reuse of the Milan Expo 2005 area proposal. 2015

G. Strappa (coordinator), P. Carlotti, I. Taci, C. Tartaglia, D. Nencini, V. Buongiorno, G. Ciotoli, M. Falsetti, A. Del Monaco, V. Matteis, P. Posocco, M. Raitano, G. Valeri, P. Marziano.

The ***central theme*** *of the design proposal is to link the different fragmented areas of the Expo held in Milan in 2005 allowing their inclusion in the existing urban structure* (Fig. 5.9)*, despite the problem of connection due to the presence of large-scale transport infrastructures.*

The **main goal** *is to rethink and transform the area of the Expo following the study of the formative phases of the territory. Green spaces serving the new university and research facilities are planned, as well as the settlement fabric, transforming and reorganizing the intersections of routes of adjacent centers into nodal urban places, capable of restoring a new coherence to this marginal area.*

Fig. 5.9 Proposal for the reuse of the Expo 2015 area in Milan. Linear settlement between the new territorial polarities of Baranzate e Pero

Fig. 5.10 Hypothesis of Roman centuriation in the north-eastern outskirts of Milan and phases of consumption of the ancient grid

Forming Process of the Urban Fabric

In the belief that reading and project basically coincide, a study of the formative phases of the area of Baranzate-Pero was undertaken (Fig. 5.11). *The soil is today predominantly flat, crossed by small waterways, sometimes lost among industrial sprawl and confusing housing expansion. We can schematize these phases through the following succession:*

First formative phase (1888):
The territory is gradually structured through the formation of small centers at the intersection of routes originated by the shape of the land and the consumption of the original ancient Roman layout.

Second formative phase (1970):
The network of large autonomous infrastructures crosses the minor centers of the peripheral area to the northwest of Milan (as is the case of Baranzate and Pero) without relating to the existing routes that guide the development of the residential and productive fabric.

Third formative phase (2016):
In the urban centers of Baranzate and Pero, it is hypothesized that routes could be "knotted" connecting different parts of the city together within a new urban organism capable, through the formation of a highly hierarchical layout, of ordering urban sprawl.

Fourth formative: design intervention
As a result of the urban structure reading of the existing routes and fabrics related to the two poles of Baranzate and Pero, a new restructuring axis connecting the two new poles is formed, marked by a strong level of organicity between routes and the newly formed fabric.

Fig. 5.11 Forming phases of the urban fabric and the new design as the last transformation phase

The proposed new constructions, traversing a dense urban "forest," pose the challenge of gathering and ordering in a single formal built gesture housing, tertiary, recreational, and cultural functions.

The building consists of two linear urban aggregates of different sizes, one fourteen stories high and the other eight. The first four levels organized in a vertical fabric consisting of duplex housing arranged in terraced and courtyard types, served by whole open galleries intersecting double and triple-height atriums to provide ventilation and lighting (Fig. 5.13). *The upper part of the lower volume is similarly organized. As for the housing sizes, they are flexible and combinable and range from 35 to 150* m^2.

Fig. 5.12 The new structure in its context

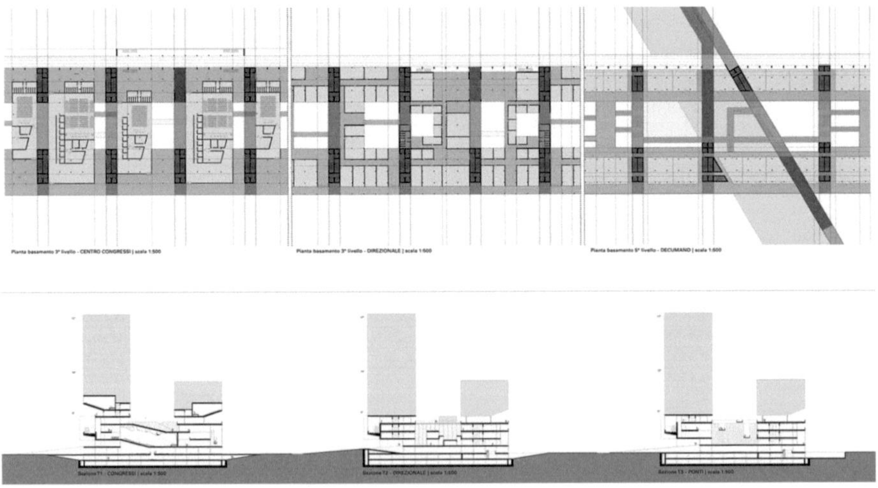

Fig. 5.13 Floor plans and their different typologies

The new linear building, in which all the required construction is concentrated, connects the two poles by overcoming the fractures caused by the transport infrastructure, and restores (compensates) this peripheral and marginal portion of the landscape its green character with the establishment of a new parks (Fig. 5.14).

"Vertical Fabric" and the different typologies.

In the urban centers of Baranzate and Pero, it is hypothesized that the fabrics will "knot" by connecting existing routes together, forming new urban spaces and specialized buildings, thus shaping the poles of the new structure. It is believed that this knotting process is the fundamental datum of the intervention: the experience of large housing interventions has, in fact, taught how linear structures of this type (the experience of Corviale in Rome is valid for all), if not strongly polarized by connections with the real built, are destined not to form a real fabric and decay in a short time.

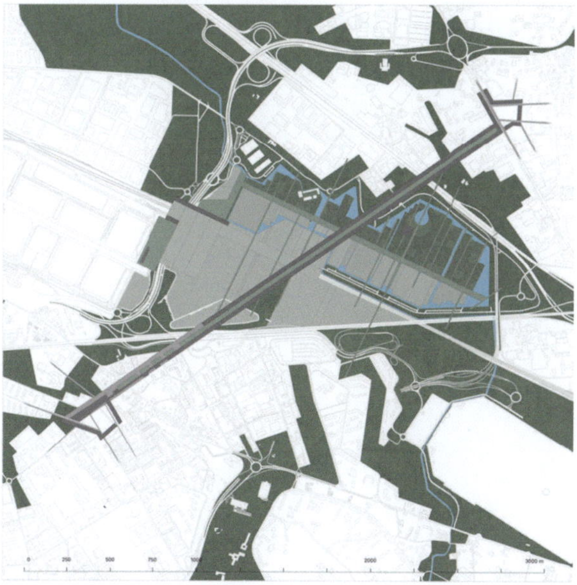

Fig. 5.14 Urban spaces and the new green areas

Project H: Urban park in the area of the former Transandina Station, Mendoza, Argentina. (selected project) 2018

G. Strappa (group leader), C.Sammarco, S. Aloisio

The central theme. *The proposal, developed within an international competition, concerns the transformation of an abandoned area in the city of Mendoza (Argentina) inside a living part of an urban organism, thus establishing new relationships of necessity with the environment at the different scales: buildings, the fabric, urban landscape. The design, derived from the reading of the forming process of the historic urban landscape, allowed the formation of a natural park integrated to the built environment* (Fig. 5.15).

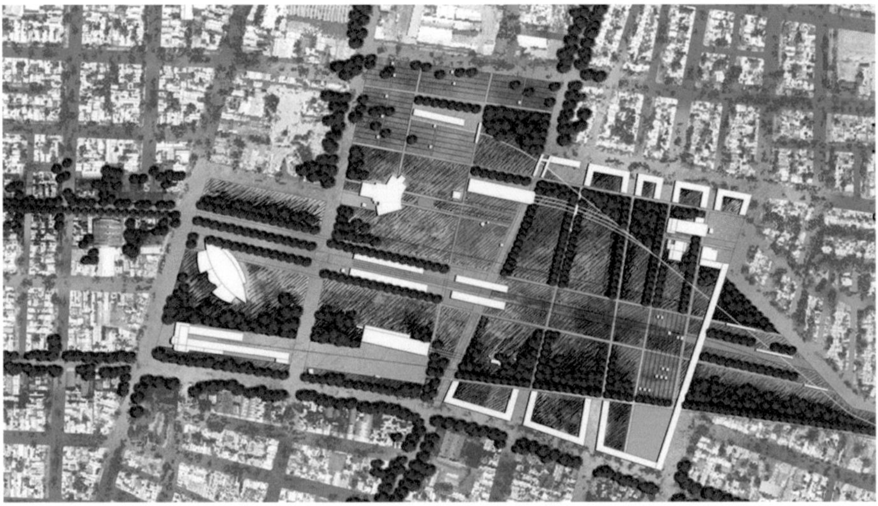

Fig. 5.15 Reuse of the Transandina Station railroad area in Mendoza. General plan

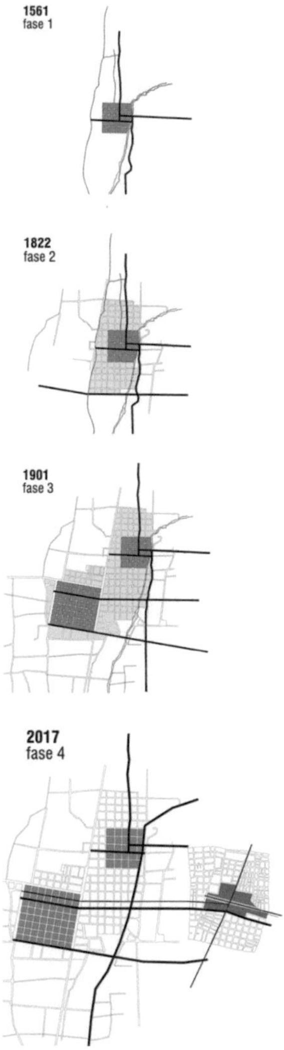

First formative phase: 1561
The first formative phase is attested to a node of territorial routes. It responds to the colonization of the territory close to the Andean Mountain chain.

Second formative phase: 1822
The occupied territory expanded in the form of a grid, resuming the routes that went from North to South, as well as the reinforcement of the original trans-Andean connection.

Third formative phase: 1901
In addition to the old center of Mendoza, a new urban center was consolidated to the northeast partially due to the earthquakes recorded at the time. Both centralities are immersed in the expansive wave of the city that sought to relate the urban environment with the eastern farmlands dedicated to the wine industry.

Fourth formative phase: 2017
The new design derives from the expansion of the fabric during the XX century. It is presented as an urban node in which the territorial routes cross, so to generate a central axis that articulates the flows in which the main functions are arranged.

Fig. 5.16 Forming process of the urban structure starting from territorial connections

Transformations. *The project is the result of the final phase of transformation of the city, on the one hand by transforming the old Belgrano station into a green node available to the community, which overflows and accompanies the axis through a linear park.*

Connections*. The north–south link along Allayme Street allows the articulation of different urban functions, like a juxtaposed structure including some local functions (neighborhood gardens, neighborhood centers, sports areas). In addition, a new shape was defined for the park due to the need to density the surroundings with housing projects. In coordination with the train station and BRT, the new interventions' will become an innovative meeting center for two parts of the city currently separated by roads* (Fig. 5.17).

The northern part*. The Joaquin Lavado Quino park is proposed as an agent of environmental awareness, where residents will find areas for self-consumption; producers and farmers, a more efficient way of managing waste; and visitors an exciting way to interact with the city. Hence, the establishment of an urban park and an Eco park for the community of inhabitants.*

Fig. 5.17 Connection between the intervention area, the planned city grids in the various stages of formation, and the new nodal points

Project I: Transformation of the south-eastern areas of Partinico–Palermo, 2013

G. Strappa (group leader), P. Carlotti, A. Giancotti, L. Reale (coordinators) A. R. Amato, C. Bernardini, F. Fava, A. Jelic (PhD students)

The method*. Like in other cases, the central theme of this project also is the notion of formative process related to the actual form of the city and its surroundings understood as derived from successive stages of formation and transformation of its urban and agricultural context.*

Considering Partinico a complex urban organization in progress, the new residential intervention should be conceived as a fabric that develops over time.

Architectural solutions*. The main goal is to consolidate and improve the peculiarities of the territory, by paying particular attention to the new district, downtown Partinico and the surrounding countryside. The strengthening of the connection axis between the urban center and the railway station area is conceived as a linear polarity that starts from the historic center, updating the existing fabric, defining a series of special buildings, and finally reaching the station hub* (Fig. 5.18)*, which is renovated taking on the role, together with other complementary services, of reference for the new urban expansion.*

Fig. 5.18 Infrastructure and urban fabric project in Partinico, Sicily

Forming process of the urban organism

Project proposal*. The general plan includes the formation of a new city quarter, which will evolve as a part of Partinico's urban organism by being included in the process of making and transforming of its urban and agricultural context* (Fig. 5.19).

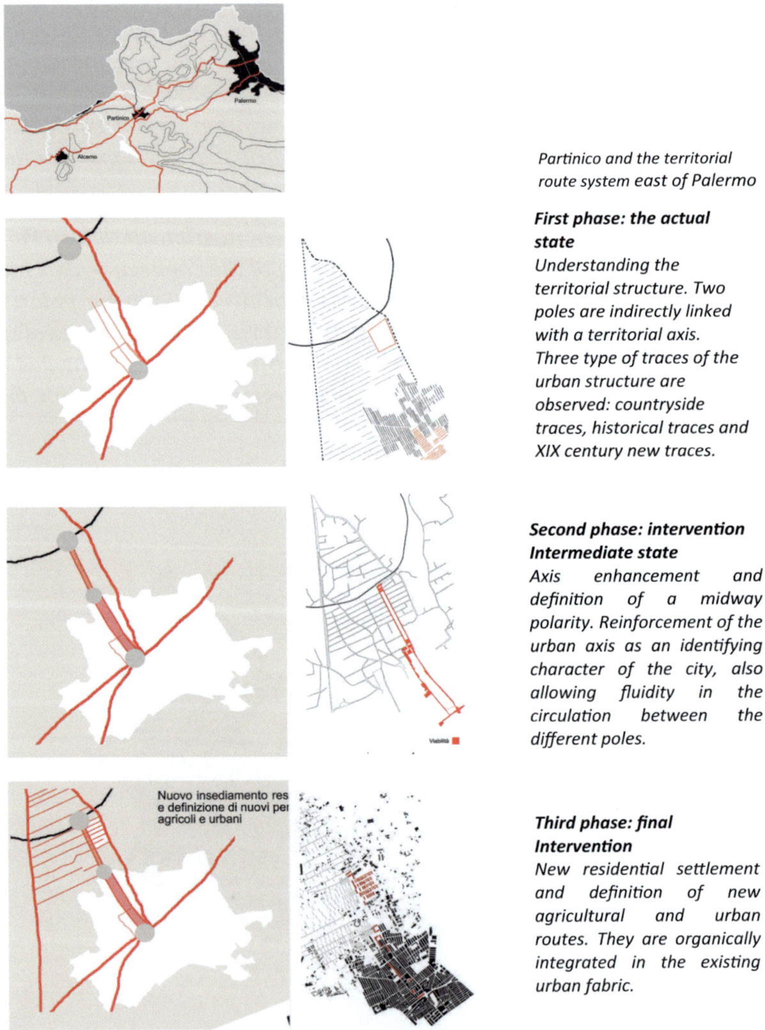

Fig. 5.19 Forming process of the urban structure starting from territorial connections up to the final intervention

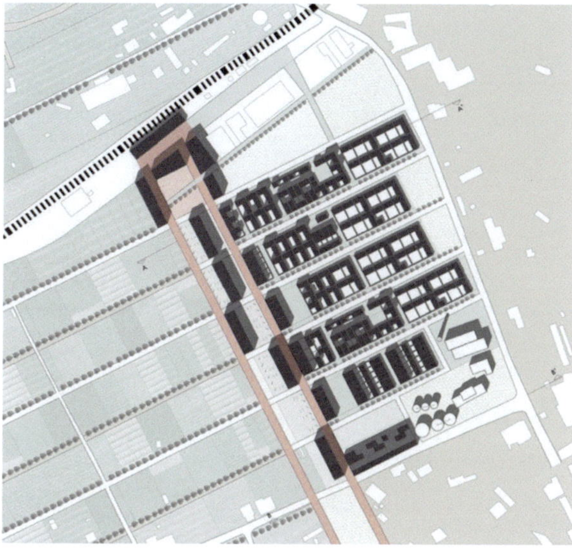

Fig. 5.20 Volumetric design of the new quarter

The central road axis*, although currently rather unstressed, is intended to reinforce the link between the described poles. The axis is doubled to consist of two distinct routes, one designated to the heavy traffic and public transport and the other reserved for the local, cycling and pedestrian viability, also allowing the formation of public spaces and specialized buildings in-between these two paths and in the proximity of urban areas.*

The secondary routes *interrupt the central axis at several points (hubs) to organize unitarily the project area. Also, the same geometrical pattern structures both the division of agricultural fields and the defining concept of the residential districts* (Fig. 5.20).

New base building forming process. *The development of residential areas is planned according to the formative process based on four constructive phases, where each stage aims at satisfying the increase of housing demands and requirements for diversity of functions, as follows a common organic development of any urban fabric. Such method enables the district's growth through densification, but without further expansion through ground consumption. The process is based on the transformation of the courtyard house type, as a housing principle already present in the area in question.*

Each of the four construction phases (Fig. 5.21) *aims at satisfying each time the increasing living and functional requirements, typical of growing space.*

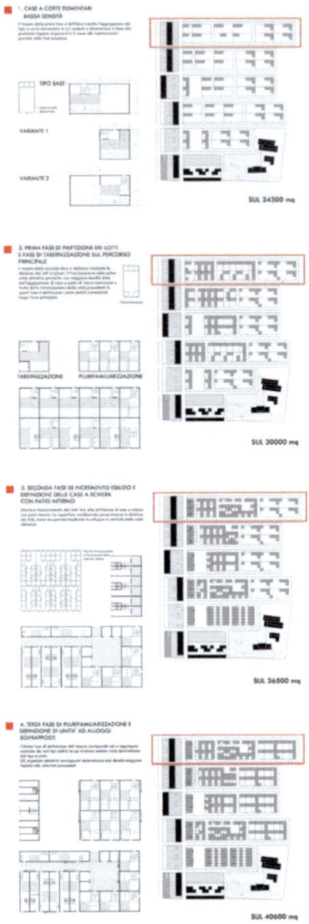

First phase: Low density elementary courtyard house
Fabric is originated from the aggregation in closed series of elementary housing units based on the notion of enclosure. From a base type derives two variants.

Second phase: Lot division
The subdivision of the first housing units allows for greater density given by the aggregation of newly built patio houses or the result of the renovation of the previous units. Also. The first commercial areas are defined along the main axis.

Third phase: Further lot division and typological definition
Terraced houses with internal patios are defined. The residential surface area lost during the division of lots is recovered through the vertical development of the housing units.

Fourth phase: Urban development
Through the use of vertical pathways (profferli), a series of overlapping dwellings are developed that are structured to avoid introspection with the pertinent areas of the units below and are capable of containing the same spatial and functional complexity as the latter.

Fig. 5.21 Future processual transformation of the housing building types

Bibliographical Note

Caniggia, G., Malfroy, S. (1986). *L'approche morphologique de la ville et du territoire*, Eidgenössiche Technische Hochschule, Zürich.

Cataldi, G., Cherici, A., Gialluca, B., Lavagnino, E., Maffei, G., Orgera, V., Vaccaro, P. (1987). *Cortona. Struttura e storia. Materiali per una conoscenza operante della città e del territorio*, L'Etruria, Cortona.

Strappa, G. (2010). *Rileggere lo Zen.* In *Paesaggio urbano* n. 5.

Strappa, G. (2018). New frontiers. Urban morphology's territories of research. In *U+D, Urbanform and Design* n.7/8.

Strappa, G. (2016). Architecture of territory. In G. Strappa, P. Carlotti, A. Camiz, *Urban morphology and historical fabrics. Contemporary design of small towns in Latium*, Gangemi, Rome.

Strappa, G. (2023). Reading the territory. Taking care of the territory. In *U+D Urbanform and Design* n. 13.

Strappa, G. (2020). Four questions to Jeremy W. R. Whitehand on urban morphology and historical cities. In *U+D, Urbanform and Design* n. 13.

Strappa, G. (ed.). (2018). *Observation on Urban Growth.* FrancoAngeli, Milan.

Strappa, G. (2016). A morphological contribution to the debate on the Milan Expo area. In *Urban Morphology* Vol. 20, n. 2.

Strappa, G. (2000). Enclosures in Islamic architecture. In *Environmental Design, Journal of the Islamic Environmental Design Research Centre* n. 20.

Strappa, G. (1998). *The notion of enclosure in the formation of special building type.* In A. Petruccioli (ed.), Typological process and design theory, AKPIA. Harvard (MA).

Strappa, G. (2019). Urban morphology and architectural design: A contemporary condition. In *Urban morphology* vol. 23, n. 1.

Whitehand, J. W. R. (1987). *The changing face of cities: A study of development cycles and urban form*, Institute of British Geographers Special Publication no. 21, Blackwell, Oxford.

Chapter 6
Expression

Abstract This chapter deals with the problem of expression that in morphology we could define as "readable order" and, at the same time, individual "aesthetic synthesis." It is, in other words, a design and compositional problem. The topic also raises the problem of language in the use of architectural and urban form. The theme of expression is today complex as the forms currently produced end up belonging not only to their own cultural pertinent area, or exchanges between areas, but to the more general content of an universal media circuit in crisis of transformation. The disorder that this condition generates should be considered a resource, the seed of transformation and innovation. Several examples of projects that deal with the problem of expression from a morphological point of view are provided.

Keywords Expression · Cultural area · Aesthetic synthesis

N.S. The topic of morphology requires, inevitably, the need to introduce the question of language, which you also define as the "legible order" of architecture. It is a very complex subject and one that, I believe, to be extremely topical when traced back to issues that pertain to "clarity," "coherence," "conciseness," and "expressive capacity" of architecture.

G.S. You are posing a composite question that would take a long time to answer. I would say, in short, that language is one of the great misunderstandings of contemporary architecture. Language is understood as a tool of individual expression when it should be, first of all, the means of communication. It is the sharing of meanings, the code, and the prerequisite that allows us to express ourselves and be understood. If this medium is not possessed by a civil community, there can be no language intended as a system of signs regulated by a conventional order. This is a vital problem for architecture today.

I would like to put it in very practical terms. Today we talk about the language of Daniel Libeskind, Steven Hall, Frank Gehry, for example, but their works do not refer to any common language of which their works can constitute the personal use. It is as if, when writing a story, we use an invented language. In fact, what we produce are images intended for a planetary market, a spectacle to be consumed.

R. Salamouni et al., *Architecture in the Making*,
https://doi.org/10.1007/978-3-031-86833-7_6

Great noise and, often, great smoke.

After all, the spirit of every protagonist seeks spectators and consensus which, let us be clear, in the conditions of crisis of contemporary culture is entirely legitimate. But language is something else.

The truth is that we have lost the art of naming things. In his anxiety to produce new and extraordinary forms the architect does not refer to any shared code.

In other words, seeing things from a diametrically opposite point of view, architecture is paradoxically today just communication, but of a banalized form, that immediacy of the advertising lexicon to which it is enough to evoke a message without any need to articulate it. After all, our profession has become a specialized sector of the global division of labor that no longer requires the study of architecture. In fact, this job presupposes a complex order that is difficult to learn, also related to construction, use, congruence with the environment, all of which must then be composed in a readable aesthetic synthesis. Precisely because this order, which once expressed the unity of things, has been lost today. I believe we need to go back to the roots of language, to its most obvious and deepest, still recognizable meanings.

If it is impossible to reconstruct the organicity of the built world, however, the need and desire are deeply human.

Which, moreover, has perhaps never existed. The organicity of the Renaissance city was a representation. It expressed a need and a desire, not the real city, which was the site of constant conflict. Order is a necessary aspiration, not a condition. Life does not allow itself to be caged by rules, no matter how much effort we make. However, those efforts are necessary in many ways: to be in the world with others, to participate through each new construction in the formation of fabrics according to common principles, and to share the beauty of an architecture in which we recognize communicable meanings. Language, as a tool for depicting the multiplicity of life, should contain the effort, difficult for any architect immersed in the contemporary condition, to overcome the limits of the individual vision of things, which often does not allow us to grasp the totality, preventing any shared synthesis.

This sharing, in my opinion, is the center of architecture and concerns both its production and its reading. It is the way in which generations of readers identify meanings and connections between things, giving recognizable sense to the transformation of the built world.

Of course, interpretations also change over time, and with them the values that architectures convey to us. Let me explain further with an example. Let us think of the different ways of reading the collaboration of elements to form the structures of architecture.

Let us think of the time of the Baroque up to Borromini, the rule-breaking consolidated myth.

From the end of the XVIII century, especially with the critical work of Francesco Milizia, through which this phase of architecture was established as an art without rules (in which the union of the parts does not adhere any longer to common principles), up to Benedetto Croce's accusation of anticlassical lack of unity.

Bruno Zevi, taking up a strand of interpretation that runs through Jacob Burckhardt, interprets it, on the contrary, as the historical premise for the modern season of freedom and creativity. Both readings mirror each other; they use, with opposite meanings and values, similar categories. I believe they are both literary interpretations, coincidentally not very different from those that determined the alternating critical fortunes of Baroque poetry. Certainly, Borromini looked at the past with new eyes; he did not want to be "just a copyist," as he writes in his *Opus Architectonicum*.

But his forms are all strictly necessary. They are subject to a general rule. In fact, Borromini is not in negation with order, but on the contrary, in obsession. His efforts transform the chaos of elements (the tumult of words) into a system ordered by a general rule that represents a universe. Like a cosmogony: the rule taken to its extreme limits, new geometries interpreting in an unknown way the bearing of history constitute the essence of Borrominian invention.

To conclude: it would certainly be very simple if architecture were a universal language, an instrument of exchange between different cultures, as internationalist rhetoric would have it.

But it is not.

The new forms of architecture are, fundamentally, related not only to the context that generates them. Above all, they are connected to areal exchanges and the disorder they generate. Which may be anything but an obstacle. They are a resource, indeed, the cause of the change that architecture must witness, document and, above all, transmit. For one of the tasks of language is precisely that of "passing on," a crucial aspect of the problem. This responsibility of actively witnessing our work is usually seen as alienated from the times, an activity aimed at the past in a society, such as the contemporary one, dominated by the hegemony of the present.

I will try to explain its relevance by interpreting some remarks by Franco Rella.

Rella distinguishes the figure of the spectator from that of the witness. For him, the spectator observes the action with detachment, as a spectacle to which he is a stranger and of which he has no responsibility. The witness, instead, is involved in the action he reports and has the task of conveying the events in which he has participated or which, I might add, he knows through other testimonies that involve him intellectually and emotionally.

Witness is the Ishmael of Moby Dick whom even the sharks respect because he is called upon to bring back to posterity the tragedy of Captain Akab and his Pequod. Similarly, through the eighteen books that remain to us of his *Histories*, the Syrian Ammianus Marcellinus is the witness of the catastrophe of the Roman Empire, of its crisis until the defeat at Adrianople, of the fragmentation of the urban organism of Rome (a house containing many cities) as a mirror of the fragmentation of the empire. Behind his seemingly detached prose, one glimpses an intense involvement, the epic and pain of a civilization discovering the coming of its own end.

But witness is also Palladio, who did not know the ancient world directly but knew of its ruins among which he wandered at length on pilgrimages to the antiquities of Rome. He touched them, recognized them, and reconstructed them. Believing he was transmitting the ancient language; he actually gave them new meanings. It is not, his, a philological operation: it is an active operation, a posthumous testimony that is also a project of resurrection, a Renaissance, precisely. Palladio "encounters" the project of a new world that gathers into unity the multiplicity of fragments surrounding it,

Fig. 6.1 Design for a church and parish services in the Tor Tre Teste quarter in Rome, 1995

just as that of Ishmael/Melville is not a simple exposition, it is a great fresco, dense with allegories, symbols, and metaphors of the condition of man lost on earth.

N.S. It would be interesting to understand how these reasonings of a typological-linguistic nature, and therefore strictly pertaining to the aesthetic-expressive dimension of architecture, have conditioned the choices of certain projects. For example, the project for the Tor Tre Teste church in Rome and the one for the City Hall of Santa Marinella.

G.S. Of course, in the contemporary condition, the task of transmitting what I was talking about is not easy. However, I think that the architect fulfills, even today, this role much more than he is willing to acknowledge.

In the projects you cite, we did not posed the problem directly, but the operation of transmitting, with the responsibility it entails, is part of our culture, also of my education as an architect and as a teacher.

The church in Tor Tre Teste was designed in 1994, together with Gian Luigi Maffei and others, on the basis of a call for proposals requesting the construction of a religious building in a suburban area of Rome, linked to anonymous urban surroundings.

It seeks to witness and convey the notion of urban knotting (Figs. 6.28 and 6.29), clearly or latently present in every special building in the city, the formation of a knot through the confluence of paths. Here, the idea is developed didactically, but the synthesis, I believe, is far from obvious (Fig. 6.1).

The logic of the routes that link the building to the context generates the architectural geometry, the formation of the central assembly space spatially nodal (Figs. 6.2 and 6.29), functionally served, statically supported, together with the layout and form of the architectural structures, to the formation of peripheral, serial, serving,

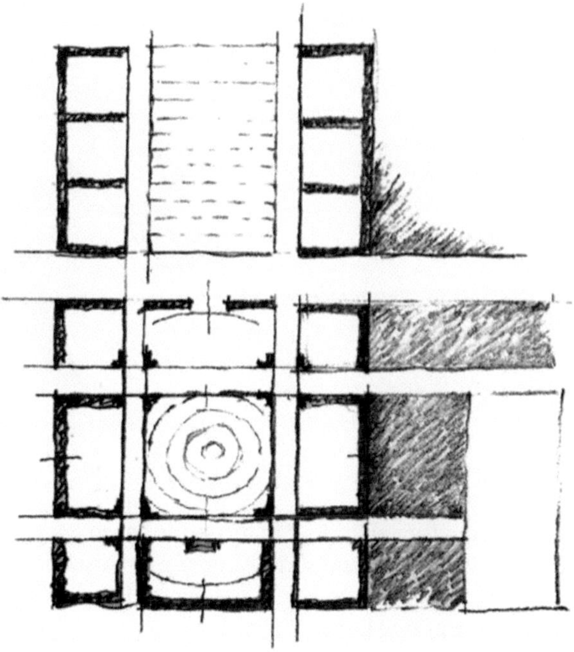

Fig. 6.2 Structure of the Tor Tre Teste church as a knotting of routes

statically supporting spaces. Special care was given to the light, coming from above, reflected from the intrados of the metal vault.

As for the new town hall in Santa Marinella, designed in 2004, it clearly conveys, I hope, the idea of urban knotting (Fig. 6.12), which is one of the most frequent themes in my work. The site evokes, for Roman architects, the suggestive presence of the La Califfa and La Saracena villas, two significant experiments by Luigi Moretti in the use of the free plan in plastic—masonry architecture. The new town hall was designed, however, to be a public space rather than a building. It needed, above all, to perimeter an open and protected place, an intersection of urban routes surrounded by functions related to the collective life of the local community. Open space, enclosed within the built, in the history of the Italian city, derives directly from the transformation of fabrics knotting (Fig. 6.13): it is a link of paths that intimately unites the monumental character in which the civic community recognizes itself, to the daily life that flows in the city. The "palazzo" is the most typical form, at the building scale, of this process of specialization knotting (Figs. 6.14 and 6.15). The new city hall is thus an open urban palace.

Of the palazzo were inherited the deep structure related to the "reversal" of urban paths within the building *knotting* (Fig. 6.16). For this reason, the new city hall was to be, above all, fabrics that specialize, linked to formative rules proper to special construction that has been updated over time while holding firm to certain generating principles.

Fig. 6.3 Study of the space for the Santa Marinella assembly hall

It was envisaged the use: of tuff in the basement part, read as load-bearing, opaque, massive; of transparent, serial structures in discrete elements employed in the elevation part; of metallic and glazed elements in covering the spatial and functional node of the assembly hall (Fig. 6.3).

N.S. I would like to ask how you express the relationship between type and morphology (that is, the changing relationship between "norm" and "form".

You approached the problem from a theoretical point of view, it seems to me, by developing a teaching method in the new architecture faculty of Bari, while experimenting, at the same time, a method of reading urban processes in the city of Trani. But how does this research translate concretely into projects? For example, in the case of the expansion/completion projects for the Terni Cemetery or in the case of the Vittorio Emanuele III Nursing Home in Trani. Two projects that, in my view, seek to, respectively: on the one hand, to reaffirm, by reinforcing it, an "urban settlement" principle applied to a place of peace and memory where the element of the "enclosure" is exalted through the compositional and constructive expedient of the "inhabited wall"; and, on the other (in the Rest Home project) to transform, renewing it, the form of the existing building through the serial aggregation of rooms around a courtyard. It seems to me a reflection on the character of the convent building as was the case in Le Corbusier's experimentation at La Tourette or, more specifically, in the works of the monk-architect Dom Hans van der Laan, who was able to reinterpret, by updating, the architecture of existing religious organisms.

Could you explain the ideational process that led to the development of those design proposals sharing, one would say, a precise design track: to investigate the potential of an already given architectural form.

G.S. These are two very different projects, but with similar formative principles. Certainly, memories of the many buildings that were part of my education, contributed to their composition. Seeing the works of Dom Hans van der Laan was an epiphany. I

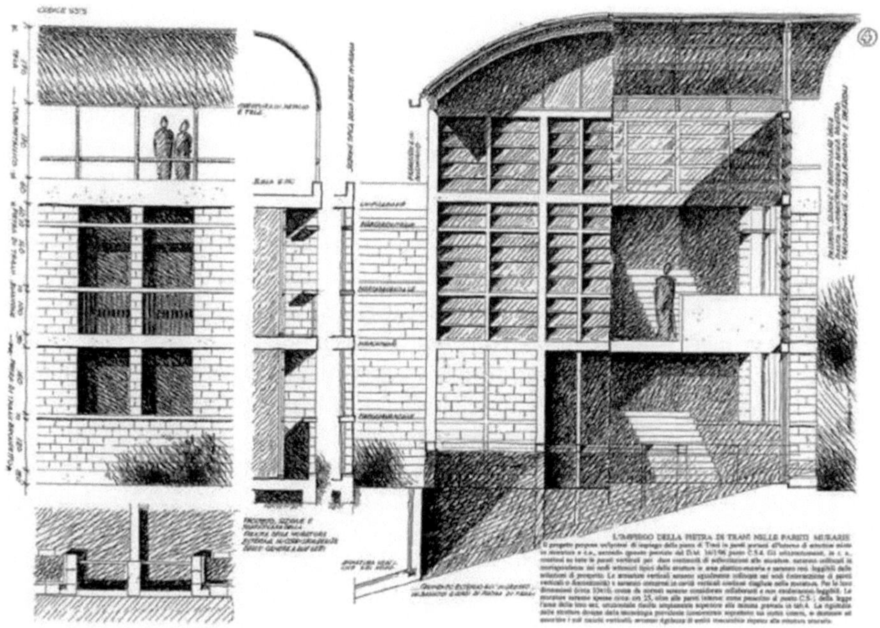

Fig. 6.4 Design for the expansion of the Elderly Home in Trani. Facade details. Theatre and common hall details

had seen his built works many years before, when they were still almost unpublished and badly worn in the cauldron of architects' repertoires. An eloquent and esoteric text at the same time open to many readings. The poetic clarity and logical simplicity of those spaces remained in my mind for a long time, even though the intellectual theoretical text, in which the Benedictine monk attempted to make scientific sense of his work (and in which I had pinned many hopes), had been, for me, a disappointment. The architecture of Van der Laan remained in my memory as a tectonic substratum, in the same way as the lectures of Louis Kahn, Mario Ridolfi, and Saverio Muratori.

The first project you ask about has to do with my work in the school of Bari. Collaborating on its foundation has been for me a fundamental laboratory of knowledge and experimentation on masonry-plastic architecture. It was not only a didactic experience.

I believe that the processual reading of Trani, published in *The City as Organism*, (Adda publisher, 2003) is a testimony to this. It is the result of a research begun, together with Matteo Ieva, then a young researcher, as part of the course of Typological and Morphological Characters of Buildings that I was teaching at the Polytechnic school.

It intended to present a method of reading based on the assumption that the different scales of reading the built environment were closely linked by the notion of organism typical of masonry/plastic areas, where each built form remains over time, modifying itself plastically (Figs. 6.4 and 6.5).

Also, we wanted to prove, how the tools of processual analysis were not subordinate but complementary to traditional tools employed for reading the city and the territory, such as urban history or archaeology.

One of these courses was the beginning of an investigation on the city of Trani (that has continued over time), including collaborations on the urban plans of the various administrations that have succeeded one another *knotting* (Fig. 6.16). On the subject, then, we published, with Marina Dimatteo a book (*La città come organismo*, Dedalo, Bari 2003). The case study was of great interest for morphological analysis precisely because there were few historical documents prior to the IX century (and none concerning the presence of an ancient substratum), and the reading therefore had to be based primarily on logical and processual interpretations *knotting* (Fig. 6.17).

The stages of formation and transformation that created the present city had to be recognized. A long comparison with the heritage of building, urban, and territorial culture common to the Apulian areas needed to be done.

We demonstrated, I believe convincingly, how the medieval structure of the city, formed first within the XI century Lombard and then XIII century Frederician wall, was derived from an older matrix, perhaps the layout of the Roman *Turenum* indicated in the *Tabula Peutingeriana*. It seemed a leap in the dark.

Once the search had been concluded in the area supposedly occupied by the forum, of capitals that had belonged to an ancient temple and used in the foundations of the first church built in the city center (that of San Martino), the subsequent discovery was to confirm, instead, the validity of the hypothesis.

The validity of the hypothesis had to be confirmed, however, by the subsequent discovery, precisely in the area that we assumed was occupied by the forum, of capitals that belonged to an ancient temple and used in the foundations of the first church built in the center of the city (that of S. Martino).

The design for the Vittorio Emanuele II nursing home in Trani is the design expression of a morphological research.

The work consisted in a rewriting: to enlarge an existing building on a very strategic area. In fact, it is in the periphery of Trani's old town, right next to the XIX century expansion. It is also situated on the route that united the city to its territory constituting the matrix of its urban fabric and establishing the legible order between building aggregates and large specialized buildings.

In the same way, as many special buildings of the past, the original convent, later transformed into today's nursing home, was in an antinodal place to the urban center, tending to assume, over time, the role of a new urban node with the expansion of the city.

A task that is not yet fully accomplished, in part because of the type of built fabric on the edge. It mainly consisted of hypothetical high-density housing, without any organic hierarchy. This is partly due to the isolation that connotes, in contemporary conditions, a retirement home.

In fact, this building types were formed according to the logic of an institution of segregation, ending up isolating those who inhabit them in spaces like those of care, interpreting the condition of the elderly as an illness.

Fig. 6.5 Perspective sketches of the Helderly Home in Trani

Starting from these considerations, the project assumed that the extension of the current structures (Fig. 6.5) would have to stand in continuity with the transformation process of the existing ones, both with respect to the urban organism, by assuming the role of a specialized center; respect to the building organism, continuing the formative process begun by the convent.

Using the opportunity that the building was to be planned in load-bearing masonry and that reinforced concrete was also to be used to resist seismic stresses, we were able to experiment with the formation of new tectonic nodes *knotting* (Fig. 6.18). They were in continuity with the masonry language of the windowsills line, lintels line, and unification band (Fig. 6.19).

N.S. The project for the Terni Cemetery seems to be based on the same idea of enclosure, but with different outcomes not only due to the scale (in this case territorial) but also to the fact that it was not designed on preexisting structures,

but on a rural area adjacent to the old cemetery, posing the problem of invention, of the design of completely new forms.

The one for Terni, you are right, is a much more complex project.

In the architectural theme of the cemetery, the failures produced by the process of specialization of the European city appear indisputable. Paragraph at the head intended primarily for the living and their memories. A poetic theme that induces to communicate to the visitor a shared idea of loss, of ending. In times gone by, members of a religious or civil community were buried *ad santos et apud ecclesiam*: they still belonged, in some way, to the life of the city through the memory of their example or passions or dramas.

Still in the second half of the XIX century, even positivist thought opposed the idea of expelling cemeteries from cities by claiming a civic role for burials. But the modern city first drove cemeteries out of the city and then reduced the problem of death to private mourning and its architecture to a mere public service. It is no coincidence that the cemetery was the only major architectural theme erased, with rare exceptions, from the research of the Modern Movement.

Instead, the design of the Terni (Araújo de Oliveira 2022) cemetery is a metaphor for a concluded and autonomous city with its own laws governing the society of elements it contains (Fig. 6.6). A city *analogous* to the real one that encompasses the problems of morphological analysis, architectural language, and also the poetic interpretation of an extraordinary theme that includes the essence of architecture.

I believe that this project can perhaps introduce the topic of the relationship between reading and design, which has always been, in history, inevitably contradictory.

It is an important issue.

Allow me a brief parenthesis.

Vitruvius, in the construction of the Basilica of Fano, did not follow the principles he had fixed. In his treatise, he wanted to indicate a way to identify, rather than forms, *generic* and *generative* types, open to different interpretations.

In the same way Saveiro Muratori, giving a contemporary example, was very un-Muratorian in his well-known intervention in the Tuscolano neighborhood in Rome. In fact, the formative process of the Roman multifamily house, which he had also long studied, seems entirely contradicted by the unprecedented architectural outcomes.

On the other hand, the best-known Italian theorists who have dealt with the problem, although they have conducted exemplary studies of the morphology of the Italian city as Aldo Rossi or Carlo Aymonino, have always clearly stated that the project does not derive so much from the reading of the built environment as from its own autonomous mechanisms. For them, the reading of cities and fabrics has its own general cognitive task; it is not a tool for designing. I am convinced, on the contrary, that the reading of the formative processes of fabrics and building types not only constitutes the essence of design itself but is part of the civic role of our profession. I believe that it is, indeed, a duty to insert what we build today in the flow of the transformations of the city and the territory.

And precisely, this method allows us to not imitate the past: we are obliged to be contemporary.

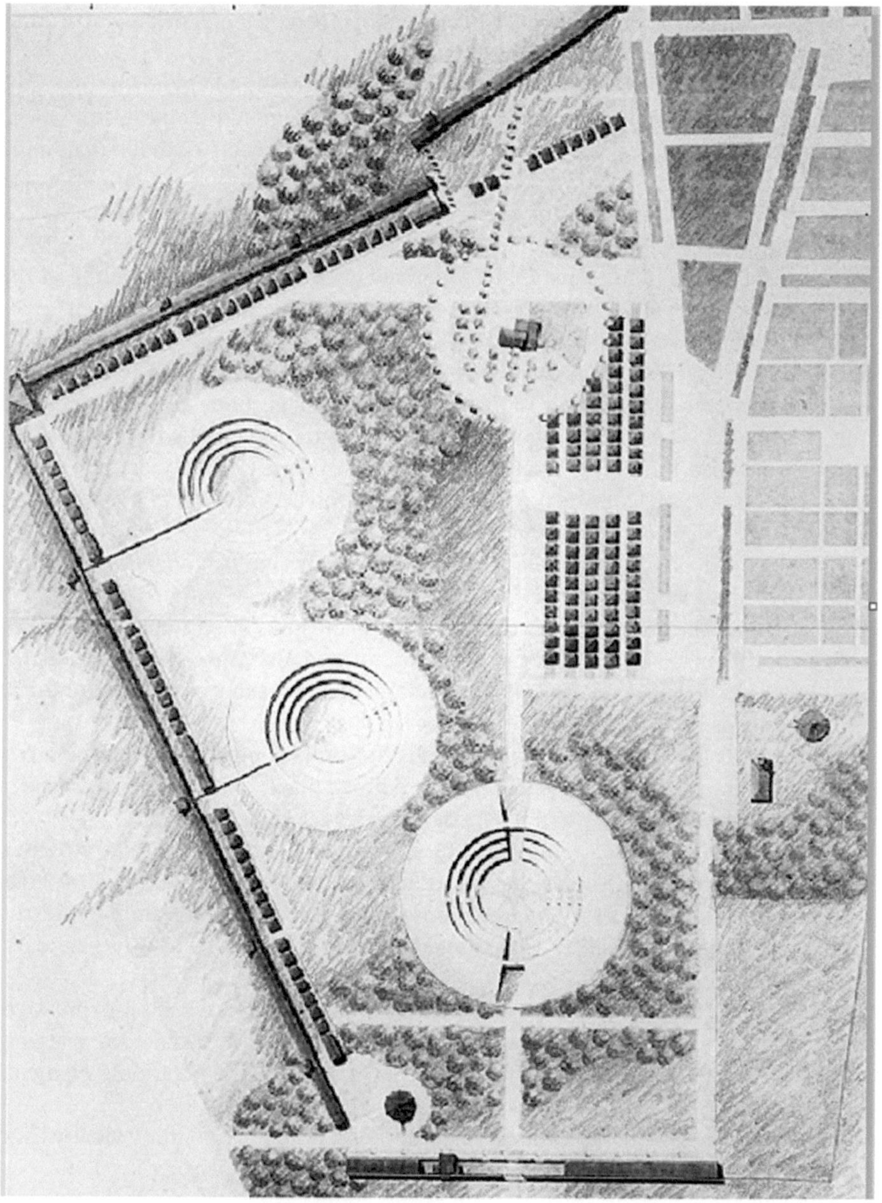

Fig. 6.6 General plan of the new Terni Cemetery, 1987

Our individual aesthetic contribution can only be personal and original. It is thus a provisional synthesis of an ongoing process.

Even if, it must be said, we have to accept contradictions and unknowns of a craft containing, despite every effort, a certain ineradicable degree of empiricism.

Recognizing some general principles, borrowed from the city of the living, in the reality of the many burial places that piety, the cult of memory or admiration for exemplary lives have sown everywhere constitutes, as well as a tool of reading, a creative act of design.

The prediction of future development derives from the innovative reading of the generative laws of past organisms.

The project is based on two dyads of complementary notions derived from the critical reading of the formative process of the modern cemetery. The notions of *enclosure* and *cover* associated with the fundamental gestures of inhabiting and protecting spaces. These notions coincide with elementary symbolic forms, meaning by this term those forms that synthetically and universally reunify certain fundamental aspects of knowledge by providing a conventional expression of them *knotting* (Fig. 6.20).

And then the dyad formed by the notions of *hierarchy* and *specialization*, linked to the previous ones, which give rise to the articulations of space within a constant and universal code. The formative processes of the enclosure, in particular, are linked to the concepts of centrality and peripherality, of nodality and antinodality which constitute a key to interpretation of the fundamental character of architectural and urban organisms.

The layout is based on the analogy of behavior of its components with the urban organism: on the idea of the cemetery as a "confined place" and, at the same time, a place that belongs to the city and its history.

For these reasons, the Terni Cemetery is articulated as an urban structure (Fig. 6.22), consisting of routes, base building, special serial building and special nodal building (the tower for congregations) (Fig. 6.21), nodes (mainly expressed by the stair towers), and poles (constituted by the intersections at a larger scale) (Fig. 6.23).

In other words, as in any organism, the rational-functional, spatial-expressive, static-constructive hierarchy of elements and structures tends to take on a legible definition through the motion that takes place in space and the polarizing elements that govern it.

One of the fundamental problems from the morphological point of view was to employ structures with a plastic-masonry character.

The topic lends itself to some general observations.

An important consideration in the choice concerns the character of organicity inherent in the use of load-bearing masonry that regulates the order of the elements (Fig. 6.7). A good masonry building is bound to physical laws that impose an organic hierarchy between the parts, yet differentiating in the case of the Terni Cemetery: the *basement and elevation*, massive, opaque, load-bearing, and enclosing spaces; the *unification band*, linear, in concrete, constituted, by the shaped beam that supports the cast-iron connecting rods; the *conclusion* of the metal roof, elastic, light, transparent, load-bearing, and non-enclosing (Figs. 6.8 and 6.9).

The use of stone posed the problem of cladding, as old as architecture (Fig. 6.24). Throughout history, the use of stone has often involved the adoption of an "indirect legibility." In fact, the formation of tectonic nodes, of constructive solutions typified in construction, has over time become code to be employed even when those nodes and solutions were no longer constructively evident. The formation and transformation of the classical order is a clear example.

A second consideration is related to the cultural area. All building in the Umbrian area possesses strong stone characters that constitute one of the most vivid factors of cultural continuity of the civil communities in this area. Few areas like the Umbrian one have carefully guarded their building heritage in our century, updated, and innovated it.

If the term had not already given rise to endless misunderstandings, one could speak of a "spontaneous consciousness," at least in part, still operating.

The example of the great architect Mario Ridolfi, who lived here for most of his life, often cited as a modern continuer of a craft linked to the refined local culture of stone, can be misleading in this sense.

In fact, it has ended up hiding from the eyes of architects, with the originality and notoriety of his research, the permanence of a deeper, collective adherence to the local root.

I believe that mine is not a positivistic interpretation of architectural language. The civilizations that produced the great codifications have always linked the material culture of the builder to the aspirations of the civic context: not so much the culture of the artist, individual, subjective, but rather the culture, shared, of the maker.

The problem needs to be addressed today in its full complexity: the modern idea of architectural organism, from Palladio to Gottfried Semper, has privileged the role of the façade as an "indirect" synthesis of the building.

But even the Albertian idea of façade did not tend to represent the constructive datum, but to express spatial contents indirectly. In this sense, Palladio's churches represent an extreme experiment in indirect legibility, where the "metaphoric" facade becomes a description of a virtual space.

In conclusion, it seems to me that the inheritance we collect as moderns from the history of architecture largely legitimizes the use of stone as a material that indicates its original load-bearing function without having a structural role.

In the case of the Terni Cemetery, with the use of *sponga* (sponge) stone, the problem is, after all, simplified. This material must inevitably be employed with a collaborating role to the stability of the building, but not with a load-bearing function, because of its mechanical properties and relative inhomogeneity. I had thus envisioned a formwork function for the exterior stone walls, entrusting this material with a dual constructive and aesthetic function. Economic reasons did not allow the full application of this principle. I believe, however, that the collaboration between stone, employed in the dual role of cladding and constructive element, and concrete, a material that can also be employed in a plastic sense, is a fertile road to follow.

Fig. 6.7 Studies for the Terni Cemetery. The perimeter wall, 1987

Fig. 6.8 Sketch of the perimeter wall of the New Terni Cemetery, 1987

Fig. 6.9 Sketch of the perimeter wall of the New Terni Cemetery, 1995

Fig. 6.10 Study sketches for the new Laurentino Cemetery in Rome, 1990

Even the Laurentino Cemetery, in Rome, had to have in some way, its own given form. Since the ancient conformation of the land was here still legible, I thought of an underground space, modifying the shape of the ground only with some paths, entrances, and light sockets that should have constituted small nodes emerging from the submerged architecture (Figs. 6.10, 6.25 and 6.26). Unfortunately, the initial idea is now legible in the first phase of construction. The execution got out of control from the public administration, beginning with the preliminary archaeological excavations that disrupted the very orography of the site with massive earth movements. Faced with extemporaneous execution, left to arbitrary and sometimes cynical decisions, I had to abandon the project and still regret it.

N.S. I would like to invite you to face an argument of a synthetic nature. In a particular way we have addressed issues related to many aspects: urban morphology, the history of the place, and construction technique.

How do you believe the aesthetic synthesis between these components should be considered?

G.S. Picking up on a thesis to which I have already mentioned, I believe that we could acknowledge to the term "architectural language" the value of personal use of a common and shared tongue, which should be understood, studied, and respected.

Put in these terms, the problem would seem solvable through a return to the foundational matrices of language, to the common and recognizable forms of our architectural heritage.

And yet the problem is much less simple. In fact, we cannot ignore the verity that a real language has been lost, that the meaning of words either consolidates with the use or vanishes.

The only form of salvation, I believe, lies precisely in the reality of architecture, in its "materiality" that differentiates it from spoken or written language.

This explains why in times of great crisis we have returned, even recently, to the material, constructive foundation of our craft: think, for example, of the research of William Morris, of Camillo Boito, of Gottfried Semper.

It seems to me that, in the current abstracted condition, a contribution to the solution of the crisis can be provided precisely by recovering from the concrete, evident aspect of architecture.

The very notion of "cultural area," moreover, in the age of globalization, is being transformed and, even though it may seem otherwise, is being extraordinarily strengthened. One only has to think of the continuity of elastic-wooden derived characters as well as transparent structures and systems employed in the high-tech and widespread forms, in areas once associated with Gothic culture.

In my opinion, in this framework, the reference, to return to the construction of the Terni Cemetery, to the specific characters of the local language, should be considered.

The Terni area, compared to the common plastic characters of the Umbro-Tuscan area, has some specificities of which I fully realized only when putting my hand to the executive project, after many visits to modern areas of the city.

In Terni, a city of industrial traditions related to steel mills, the irruption of modern metal technologies has produced remarkable innovations. Because of these reflections, the fourth aspect of architectural layering, the conclusion that overlays the plastic-masonry structures, is, precisely, elastic in character.

In this regard, allow me to make a final remark concerning the individual interpretation of the theme.

In the construction of the Terni Cemetery, the transformations have gone on for a quarter of a century, with continuous, small but important rethinks, innovations, and updates. And yet, I never believed that reworking on the initial design was essential.

The answers to our problems, in architecture, have largely already been given. Recognizing them in contemporary terms, in the age of praising uncertainty and fragmentation, I think is a real form of innovative writing. After so many years of familiarity with their forms, I can no longer assess the emotions that these constructions can provoke anymore. I do not feel that the work conveys the melancholy of detachment, nor the pain of lost affections. Rather, it is perceived, I hope, as a calm architectural *pietas* understood in the classical sense of the term. In other words, I think it is a project that tries to avoid the romantic impulses to which might be induced by the theme. Work is continuing today with the construction of a new pavilion of extreme seriality and will continue in the coming years. The temptation of change, of keeping "up with the times," is strong. But there is no need for it. I talked about it, at the time, with Sandro Anselmi and that confirmed to me the idea of continuing the great enclosure, the long slot of shadow of the arcade, perhaps with other means, but with the same end, with the same unifying rhythm.

Project J: Competition for the new Town Hall of Santa Marinella (Rome), 2004

G. Strappa (group leader), T. Casatelli, A. Franchetti Pardo, F. Maraglioli, A. Schietroma, M.L. Tavella

***The central theme** of this proposal is to design a new town hall, acting rather like a public space than a building* (Fig. 6.11). *A place where the community identifies itself.*

The proposal derived from the idea of urban knotting. The process of specialization of an area of S. Marinella led to the design of an "open urban palace." The palazzo is, in fact, the most typical form of this specialization process of urban fabrics. In the modern Italian tradition, many new building types are derived from the structure of the palace.

***The new town hall design** intends to collect and perimeter a space, open and protected, surrounded by the functions related to the collective life of the community* (Fig. 6.12). *In the Italian tradition, these elements largely also coincide with the square structure. The chosen building type is that of the central space bordered by the built-up area, which, in the history of Italian architecture, derives directly from the transformation of fabrics closely linking the monumental character in which the civic community is recognized, to the daily life that flows in the city.*

Fig. 6.11 Sketches for the new Santa Marinella town hall (2004). On the left the new square as an urban node, on the right the assembly hall as a building node

Fig. 6.12 New City Hall of S. Marinella. Aerial view of the new intervention inside urban fabric

The forming process

Arising of the urban knotting. Design formative phases in analogy with an urban fabric formation (Fig. 6.13).

The design method. *The new city hall is linked to formative laws of Italian specialized building, updated over time while keeping the generating principles firm:*

1. *The unifying function of the routes and their hierarchy according to the urban role they are called upon to play. These routes, whose role is emphasized by the pavements, are closely related to the built fabric surrounding the project area:*
2. *the crossing route, the pedestrian routes, the corridors* (Fig. 6.13).
3. *The polarizations and nodalities that have the task of generating and giving direction to routes. They coincide with hierarchized spaces and specialized elements (structural, functional, and esthetic).*
4. *The distinction into serial elements (virtually repeatable compartments of modular size, in which the normal activities of building life are carried out) and organic elements, in which special activities are carried out.*
5. *The organization of external legibility in bands of "architectural layering" that employs a langue that is still recognizable despite the extensive crisis conditions of contemporary architecture: basement, elevation, unification, and conclusion.*

Elements and composition. *The main elements constituting the project are: the multipurpose council hall* Fig. 6.15, *the serial technical and administrative services rooms, the public court/square, and the tower. The entire organism has its node in the council hall, which represents, both spatially and symbolically, the central element of the organism.*

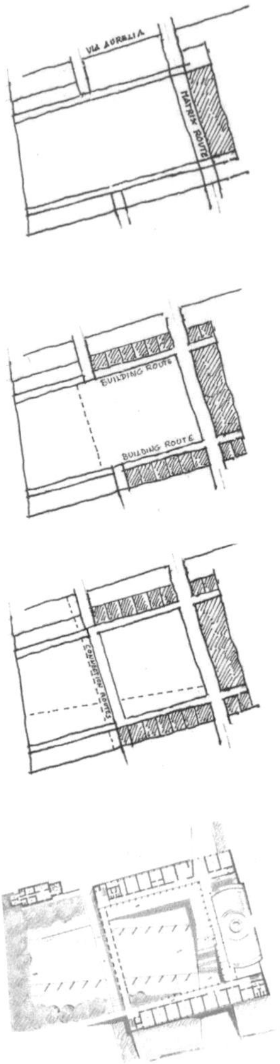

First formative phase
Formation of a special area on the new matrix route connecting Via Aurelia (territorial route from Rome) and Via della Libertà (route parallel to the coast). The process corresponds to the formation of a Roman palace at urban scale, where the private courtyard is transformed into a public space. In fact, the Roman palace originates from the recasting of row-houses with the reversal of routes within the urban routes.

Second formative phase
Formation of two new building routes on which the cells of the new urban organism are implanted, similar to the formation of urban fabrics

Third formative phase
Formation of the connecting route. Conclusion of the Internal routes structure. Last linking stage concludes the knotting process with the formation of a large internal court/plaza.

Fourth formative phase
Project design. The general character of the formative process becomes specific, individual. In particular, the project addresses the problem of height differences that articulate the matrix route and that of architectural expression understood as estetic synthesis of a formative process. The connecting route is actually double: at ground and at first floor.

Fig. 6.13 Forming process of the new design in analogy with an urban fabric

All the different spaces are connected with pedestrian routes. The pedestrian circulation will have the new square at the center of a system of paths at different heights that will make it permeable from all sides and define it as the true node of the city's thoroughfares. The system of walkways therefore represents the datum on which the overall organization of the civic center is based and constitutes, in analogy with urban fabrics, its compositional genesis.

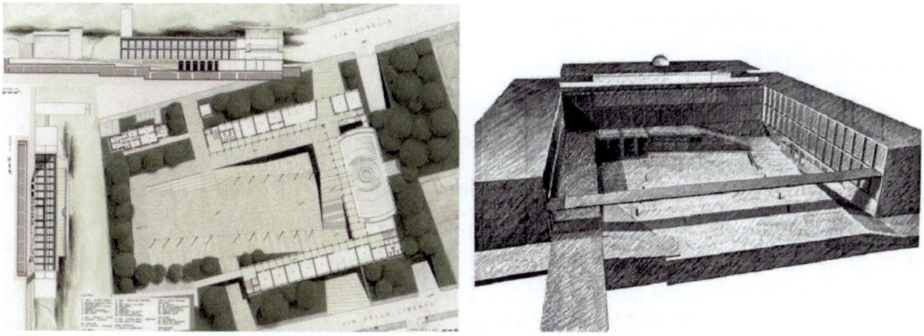

Fig. 6.14 Ground floor and perspective view of the new urban node

Fig. 6.15 Interior space of the assembly hall

Project K: Elderly home Vittorio Emanuele II new expansion in Trani, 2005

G. Strappa (group leader), M. Ieva, C. Robbe

The purpose of the architectural competition *was to enlarge an existing structure located in a strategic urban area. The extension of the current structure should stand in continuity with the transformation process of the existing one* (Fig. 6.16). *It required respecting the urban and the building organism, continuing the forming process begun by the old convent of Nostra Signora della Grazia, built at the end of XVI century.*

The design central theme *is thus the demonstrative intention to present a method of reading based on the assumption that the different scales of reading the built environment are closely linked by the notion of organism.*

Design forming process

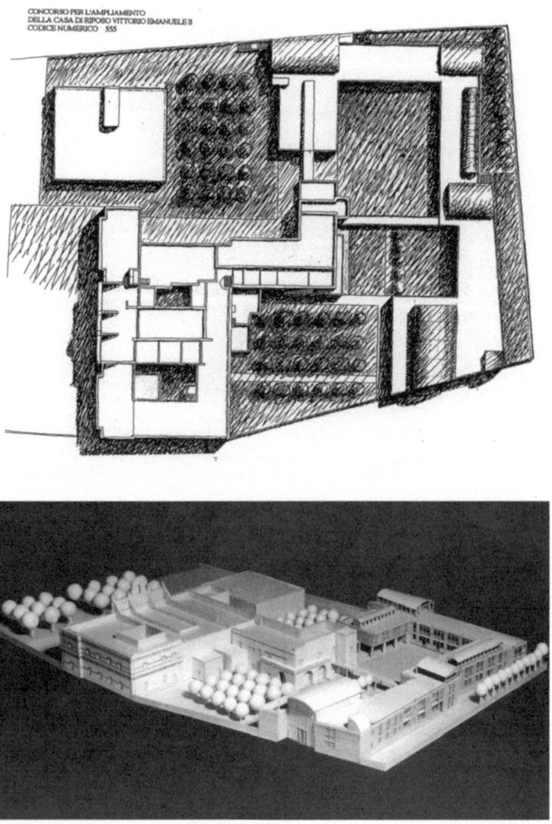

Fig. 6.16 Model of the design proposal intended as redesign of an urban portion of Trani

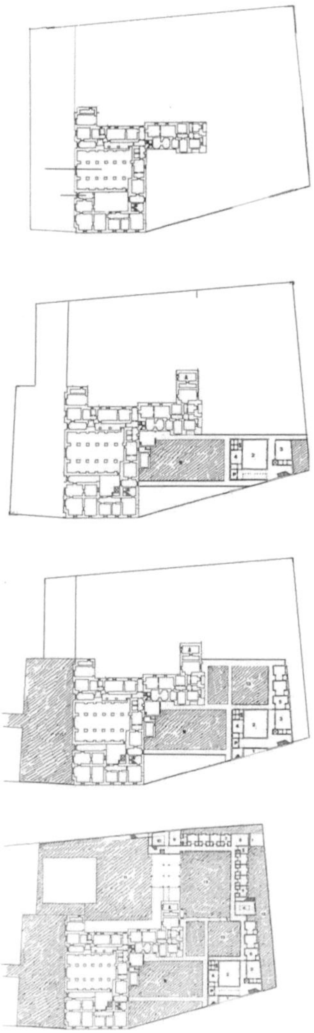

First formative phase
Construction of the small convent attached to the church of Nostra Signora della Grazia (1595). Given its small size, the convent does not have the route hierarchy features of other similar larger structures. However, to the right of the church there is access to a small corridor concluded by a cloister. corresponding, morphologically, to the connecting route formed in the last phase of the major structures of this type.

Second formative phase
Formation of the special building of the hall for common activities, meetings and performances as a knotting of the route coming from the connecting route (see the previous phase) and another one coming from the church vestry now converted into spaces for the nursing home. Other complementary spaces are assumed to be used as a daycare center with the utilization of some existing restructured rooms.

Third formative phase
Continuation of the route structure formation started in the previous phase with the placement of the first rooms for hosts and an outdoor space for open-air activities.

Fourth formative phase
Final design. Formation of routes on which the new serial rooms for hosts are implanted, polarized at the end by the mess hall, with the hierarchical spaces of the reading and living rooms.
The routes join to generate a closed structure thus forming a courtyard open to the small orange grove.

Fig. 6.17 Forming process of the new design of the nursing home

The area of intervention*. The site on which the Vittorio Emanuele II nursing home stands is located on the periphery of Trani's old town, on the edge of the XIX century expansion. It is also located on the route that not only connected the city to its territory, but also formed the matrix of its new urban fabric, establishing the order of relationships between building aggregates and large specialized structures. Like many similar specialized buildings of the past, the original convent, later transformed into the present nursing house, was in an antinodal position with respect to the urban center, tending to assume, over time, the role of urban node as the city expanded.*

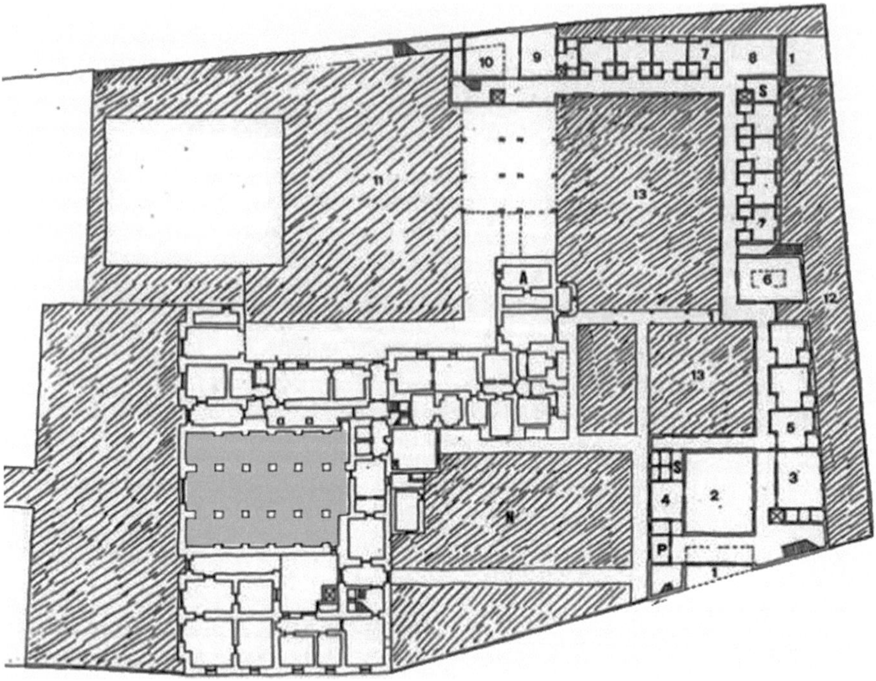

Fig. 6.18 Ground floor plan (left); First floor plan (right)

Character of the proposal. *The central theme of the project is the assumption that the expansion of the current structures should stand in continuity with the historical transformation process of the existing building:*

1—with respect to the urban organism, summing up the role of a specialized area;

2—with respect to the building organism, continuing the formative process begun by the convent

The layout. *The perimetral structure around open spaces meets the needs of a home for elderly and is widely preferable to more concentrated layouts. The new constructions continue a convent-like design, based on a logic similar to that on which the building aggregates are founded, borrowing from these the hierarchy of the corridors, the aggregation, and the hierarchization of the rooms. The serial spaces are arranged along the perimeter and served by paths polarized by specialized areas that contain common activities and exchanges with the outside.*

Construction. *The building was to be planned in load-bearing masonry and reinforced concrete. Trani's stone, due to its high hardness and resistance, was retained suitable for use as load-bearing masonry. An experimentation in the formation of new reinforced concrete tectonic nodes was done. They were designed in continuity*

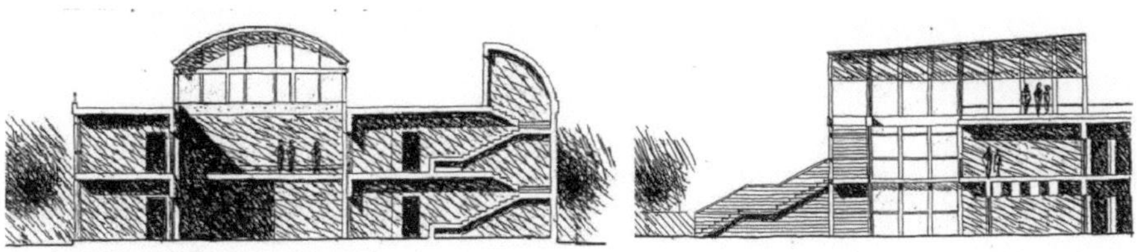

Fig. 6.19 Sections on the service spaces

with the masonry language of the windowsills line, lintels line, and unification band, ensuring the constructive continuity necessary to resist seismic shocks.

Project L: Terni's New Cemetery Expansion (second and third construction phase, 2005–2024)

(National competition first prize winner)
G. Strappa (group leader), T. Casatelli, P. Di Giuliomaria, E. Timpani

The Central theme *derives from the idea that the design of cemeteries should follow some general principles which, in architecture as in every cognitive and technical activity of man, can be expressed through symbolic forms, which constitute an important part of the inventive substance of architecture* (Fig. 6.20).

For this reason, the language employed derives from the reading of architectural types that the history of burials has sedimented over time.

Forming process of the Terni cemetery expansion design

The forming process of the project follows the logical phases of structuring a city concluded by the perimeter of the urban walls which hierarchizes the routes and internal spaces and gives order to the buildings (Fig. 6.21). *The Terni's Cemetery expansion constitutes, in fact, an organism, a set of routes related to each other by a relationship of necessity and collaborating for the same goal. One of the challenges was to identify the characters of this organism as the synthesis and conclusion of a continuous process of the transformation.*

Design method. *Terni's new cemetery is the outcome of a dyad of opposite and complementary terms: it is a piece of the city as well as a protected garden, summarizing the characters of city's great urban history with the traces of its own rural past. Logical and historical affinities are based on the aggregations of serial elements unified by routes and constituting a built organism. The reading of the formative processes provided valuable insights into the character of the intervention.*

Fig. 6.20 Google view of the Terni Cemetery with the introduction of the crater not yet built

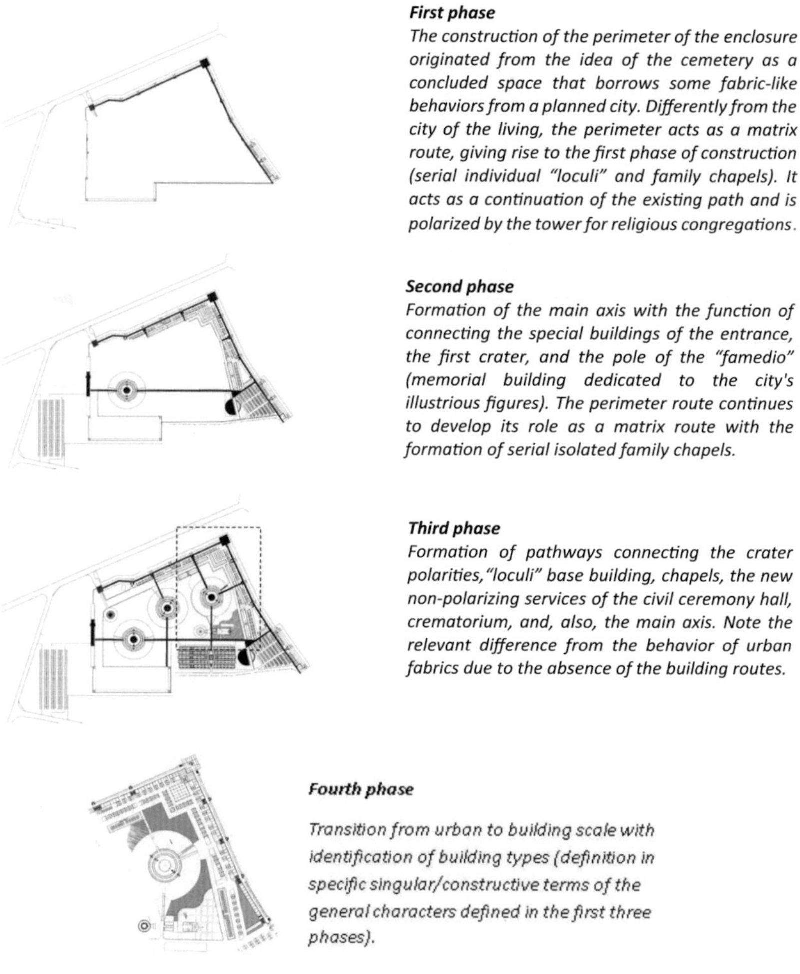

Fig. 6.21 Forming process of the Terni Cemetery expansion design

Building type and fabric. *The project originates from a reflection on the typicality of architectural forms, in the conviction that the design of burial places should follow certain general principles expressible through architecture* (Fig. 6.23). *The "inhabited" wall is a "primary archetype" that gives logic strength to the project, charging it with a powerful symbolic value and also preserving the continuity of the language during the different stages of construction.*

About this intervention Franco Purini wrote: "Giuseppe Strappa's architecture is structural, evident in the formation and process that engendered it, profoundly real, in an objective way... Realized through two phases, this important intervention has been able to self-stratify, that is, to grow on its own history by giving rise to a contained series of linguistic variations, (…), while consolidating certain symbolic

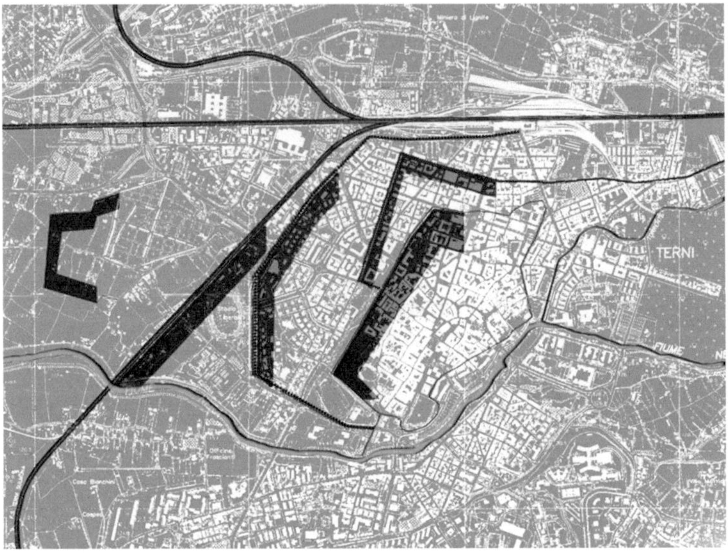

Fig. 6.22 Origin of the new form. Subsequent fortifications and urban perimeters from the Roman walls to railway line. The cemetery establishes the new perimeter between city and countryside

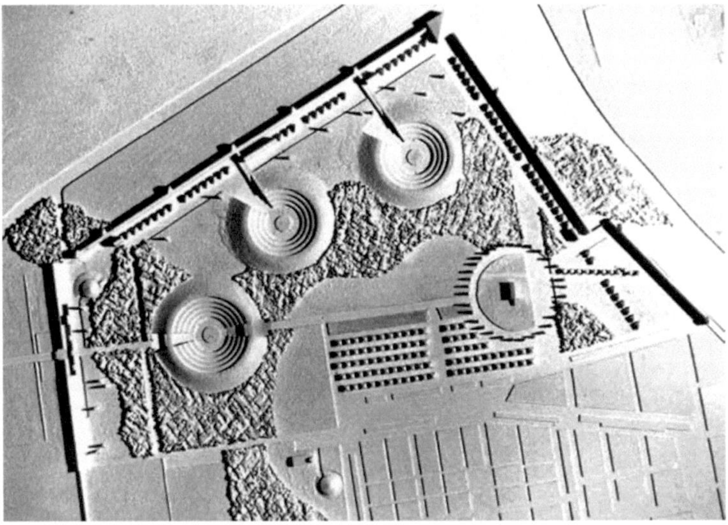

Fig. 6.23 Competition for a new extension of the Terni Cemetery, model, 1987

Fig. 6.24 Internal and external views of the ossuary tower (top); internal view and new pavilion under construction (bottom)

values ...The language of this work expresses an original moment of building, in which the tectonics stand as a logical principle that bind technique to form...." (Purini 2011).

The construction. *The building consists of a porticoed ground floor* (Figs. 6.24 and 6.25) *with family tombs, the associated burial niches of which are arranged in the basement (not accessible by the public), and a second floor of above-ground loculi (burial niches). The stone facades are marked by tectonic nodes typical of plastic-masonry structures. The stair towers are made of steel girders and brick infill, arranged for ventilation and lighting of the interior.*

Contrary to the local tradition of building individual family chapels, in aesthetic competition with each other, the area next to the existing cemetery's niches was made of chapels identical to each other (Fig. 6.26), *in reinforced concrete structure and four-pitch roof covered with tiles. The exterior walls of each chapel are of local spongy stone and courses of bricks marking the lintel node laid in a pattern.*

Materials. *The choice of using stone structures in the project is linked to the notion of "cultural area." All construction in the Umbrian area possesses strongly masonry-plastic characters that constitute one of the most vivid factors of the cultural continuity of the civil communities that have transformed this territory over the centuries. Components that, to be historically conflicting with each other, were the synthetic matrix of the "invention" (intended in the etymological sense of the term: a "found"*

Fig. 6.25 New pavilion, 2023

solution) discovered in the fruitful forming process, consisting of encounters and contradictions, that generates the form of the territory and buildings.

Fig. 6.26 The chapel from inside

Project M: Church and Parish services in the Tor Tre Teste area, Rome. 1995

G. Strappa (group leader), G.L.Maffei, T. Casatelli, P. Di Giuliomaria, A. Trombetta

Theme of the project. *The national competition asked for a religious building in a very dense neighborhood on the outskirts of Rome almost devoid of meeting places. The competition requests demonstrated how the urban magma of the suburbs, with its complexity and problems, constitutes a true laboratory of contemporary architectural.*

The new construction proposed by the team is an architectural "knot" that derives from the intersection of urban routes entering the building and transformed into internal paths, both ritual and architectural (Fig. 6.27) .

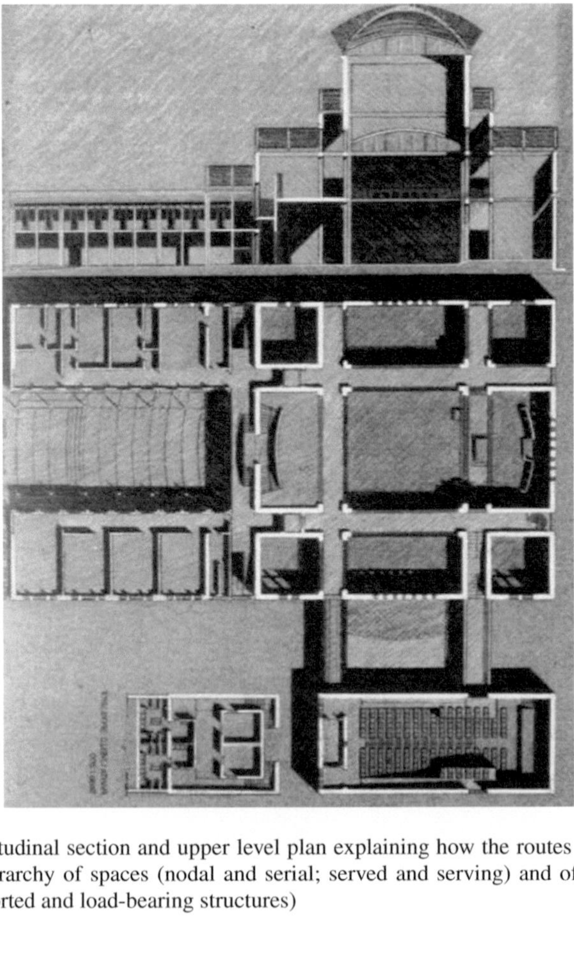

Fig. 6.27 Longitudinal section and upper level plan explaining how the routes generate, in their knotting, the hierarchy of spaces (nodal and serial; served and serving) and of the construction structures (supported and load-bearing structures)

Liturgy . *In fact, the life of the religious building consists in the ritualization of gestures of faith, codified in the liturgy, where the individual believer communicates, in forms that are continuously renewed and updated over time, the need to share his own sentiment of the sacred with the assembly of the faithful that gathers in a common space.*

The architectural "invention" of religious space consists in this supportive link between the form and life of the building (in the relationship of necessity that is established between the needs of worship and the character of the construction). The innovative design, therefore, should be considered not as a simple search for novelty, but a deep understanding of the spirituality of the architectural work.

Knotting process. *The term "innovation" brings us back to its meaning of considering which, among the infinite possibilities of change that inherited forms possess, is the one that conforms to the current condition: pertinent to its own historical phase and, even in a globalized world, to its own cultural area.*

In this case, therefore, the innovation was born as a knotting of routes that hierarchize the elements and are transformed into symbolic paths through light and the vertical expansion of spaces. This link between the ritual manifestation of faith and the physical space of worship is the center of the design (Fig. 6.28).

Fig. 6.28 Ground floor level

***Forming process of Tor Tre Teste Church Architecture*:**

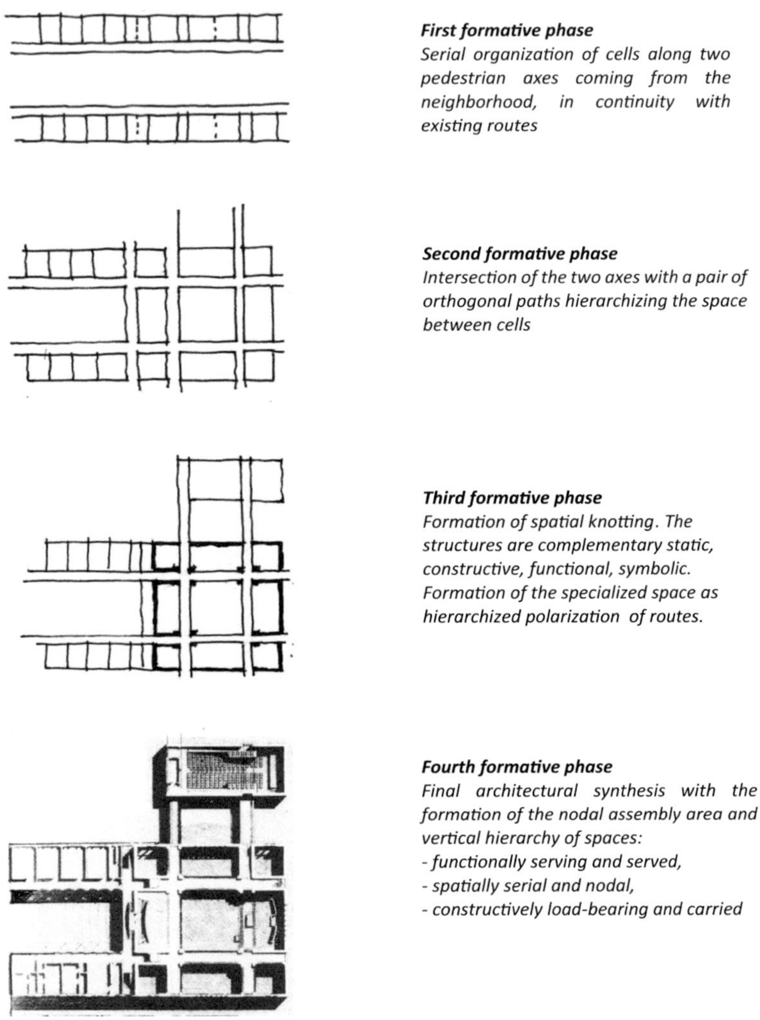

Fig. 6.29 Forming process of the Tor Tre Teste church design

Bibliographical Note

Anselmi, A. (2012). Un'architettura dell'ombra. Il nuovo Cimitero di Terni di Giuseppe Strappa. In *Paesaggio Urbano*, n. 21.

Araújo de Oliveira, V.M. (2022). Urban morphology. An introduction to the study of the physical form of cities. Springer.

Cataldi, G., Maffei, G. L., Vaccaro, P. (2002). Saverio Muratori and the Italian school of planning typology. In *Urban Morphology*, Vol. 6, n. 1.

Etlin, R. (1984). The *architecture of death.* MIT Press, Cambridge (MA).

Purini, F. (2011). Un'architettura tra origine e inizio. In AR, Vol. XLVI, n. 95/11.

Strappa, G. (1989). I cimiteri, in P.Carbonara, *Architettura pratica*, UTET, Turin.

Strappa, G. (2021). The Terni cemetery. Considerations on the relationship between reading and design. In Oliveira, V. (ed.), *Morphological research in planning, urban design and architecture.* Springer, Cham.

Strappa, G. (2022). Aree culturali e crisi della globalizzazione. In *U+D, Urbanform and Design*, n. 17/18.

Strappa, G. (2015). *Rome urban form in transformation, in City as a process, Proceeding of the 22° ISUF conference*, U+D Press, Rome.

Strappa, G. (2016). The design of knots. In G. Strappa, G., Carlotti, P., Camiz , A. (eds.). In *Urban morphology and historical fabrics. Contemporary design of small towns in Latium*, Gangemi, Rome.

Strappa, G. (2018). Reading the built environment as a design method. In V. Oliveira (ed.), *Teaching urban morphology.* Springer, Cham

Strappa, G. (Ed.). (2018). *Observation on urban growth.* FrancoAngeli, Milan.

Strappa, G. (2018). Substrata. Morphology of the ancient city behind its ruins. In *U+D, Urbanform and Design*, n. 9/10.

Strappa, G. (2019). The knotting as a morphological phenomenon. An interpretation of the Italian chamber of deputies forming process. *ICONARP, International Journal of Architecture & Planning*, n. 7 (Special Issue).

Strappa, G. (2020). Assemblage and aggregation. Reading the ancient city and urban composition methods. In *Urban Morphology*, Vol. 24, n. 2.

Chapter 7
Didactic

Abstract This chapter addresses the problem of the designer's responsibility for his own choices and the way in which its request should be transmitted to new generations. It deals with the problem of ethics in architecture and how it could be possible to overcome the de-responsibility induced by homologated forms consumprtion, which end up constituting a reassuring repertoire. It is therefore necessary to study the deep transformations substratum of the built environment which explains why things are as they appear. Studying the formative processes of the built reality, the different changing phases, the project should be considered as the critical, last phase of transformation, that of the near future. The reading of the transformation processes does not provide a solution but offers a series of congruent possibilities from which the designer has the responsibility to choose and which he is required to express through a personal aesthetic synthesis. The reading of the built reality cannot be, in other words, a description, but a vital and operative interpretation.

Keywords Formative process · Responsibility · Repertoire

N.S. I would like now to dig into the merits of how you conceive teaching activities. Your teaching seems oriented toward the search for the origin and becoming of form, of the continuous implications between things, of "signifiers," and, in a broader sense, of an operative method oriented to the transmissibility of the contents of the forms found. It would be interesting to understand how you constructed this method of teaching.

G.S. In my opinion one of the most relevant problem in educating architects is teaching them responsibility: teaching, firstly, to seek your own truth, thinking with your own head, and then sharing your ideas.

Architecture is a condition of permanent change, sometimes very slow, sometimes sudden, taking the form of a rupture (a crisis) which involves the task of biased, oriented interpretation of the built world. Yet, this responsibility is not taught to the architecture students. Someone else will take responsibility for them. The important thing, it seems, is that the students learn to stay inside the circuit of consumption.

R. Salamouni et al., *Architecture in the Making*,
https://doi.org/10.1007/978-3-031-86833-7_7

They employ forms that are supposedly new, but are, in fact, part of a global repertoire. The reassuring forms of officially sanctioned art (in which there is no question for explanation, awareness, demonstration, reason or proof) are part of a teaching bureaucracy that often spoiled the education of our students. The only necessary condition, it seems, is that the forms employed are part of an international production recognized as contemporary and legitimized by the media.

If a student presents a house on tilted pillars with no reason given, no explanation is asked of it (ratified as it is by hundreds of examples on media circuits). While if he presents a house with elementary load-bearing walls and a simple roof, he is considered a lunatic.

Even the most critically minded students accept that the magnificence of the wall, its beauty, its poetry, has become an intolerable blasphemy.

Let me tell you a little story that Claudio d'Amato once told me on this subject. A Chinese village was plagued by a long drought. A wise man predicted that rains would soon come, but that they would be a disaster: those who will drink that water would become foolish. The wise man then warned the inhabitants who, however, did not want to listen. They drank the water and became foolish. The wise man had therefore become the outsider, the outcast because he was different. Life had become impossible for him, and, in the end, he drank that water, too. This is what happens, I believe, with our best students.

Claudio has written an extraordinary and out-of-the-box book on the problems of teaching in architectural schools (D'Amato, 2015). A book soon indexed by academia, as in the story of the village where people were afraid of the truth.

Coming back to your question, the problem of teaching a reading/design method arose, for me, when, still very young, I was suddenly given to teach a design course due to the momentary unavailability of the official lecturer. Teaching was a formidable means of learning. It forced me to be clear and then to clarify, systematize, and demonstrate the directions to be given to the students. I had chosen, as the project theme of the year, the reconstruction of an area in downtown Subiaco (an historic settlement not far from Rome) destroyed by bombings.

I quickly realized the responsibility of giving students consistent and demonstrable directions, which could not but be simply related to the form of the preexistences or to functional problems.

I also understood, due to my education being based on principles inherited from the Modern Movement, how this relation could not be of simple affinity, much less of imitation. With considerable effort, and with several mistakes, I realized that the contemporary project had to arise from that part of the still valid principles, reinterpreted, updated, and innovated, that had led to the construction of the city, to its beauty of parts working together.

The notion of organism and process was a difficult achievement, obtained by trying to interpret the formation of routes and buildings (Fig. 7.1). Noting how existed, with evidence, a deep substratum of transformations of the built environment which explained why things were as they appeared in front of our eyes. Even then, there was the danger of misunderstanding. The students thought, at first, of a reconstruction "à l'identique." Only after much effort did they come to understand

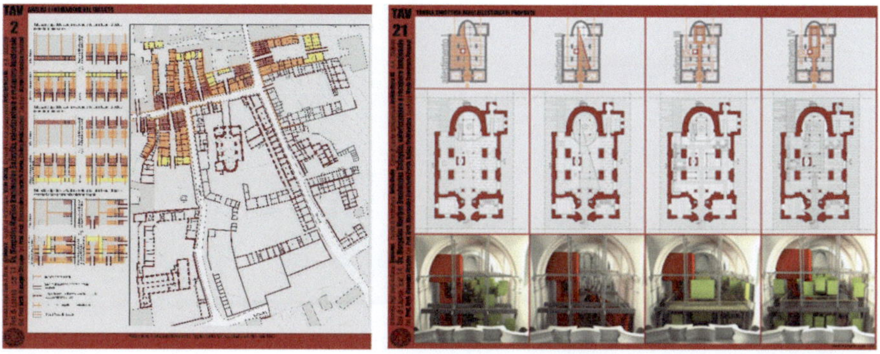

Fig. 7.1 Graduate Laboratory, academic year 2008/2009, Sapienza, University of Rome. Study of the formation of the urban fabric in the area of the Sv. Mergèles Marijos convent in Vilnius and proposal for the reuse of the church as flexible exhibition structures with the preservation of traces of the interventions in the Soviet period

how the reproduction of the preexisting building was impossible, how we could not but be contemporary.

Although uncertain, they learned to design basing on the reading of formative structures instead of imitating historical building. The course had been entrusted to me out of the blue. Under the urgency of preparing the lectures, I reflected on some important texts by Giovannoni, Milani, Muratori, and Caniggia. Finalizing the study to the design, I felt like I was reading them for the first time.

I understood in a few months one fundamental thing: how, for the architect, theory arises from practice, from the experience of designing, from seeking, in the tumult of the particular, the general law that gives meaning to individual phenomena that gives sense even to the exception, even to the fragment.

But the real test was my experience in the Polytechnic School of Bari, where I had been offered, a few days before the courses opened, the annual (and initially optional) teaching of *Typological and Morphological Characters of Architecture* (Figs. 7.2, 7.3 and 7.4).

The beginning was very hard. I would prepare topics during the week and conduct the quite challenging, four-hour long lectures in the frigid lecture hall of the Santa Scolastica monastery packed with students and with Claudio d'Amato, dean of the faculty, seated in the front row. He continually asked me, very embarrassed, questions and often seemed unconvinced by the answers. He encouraged me, however, to record the lectures and to write a text on the subject. It was very useful.

Unità dell'organismo architettonico (Unity of the architectural organism) became, over time, a reference text for the entire faculty, appreciated by students despite the prejudices of some professors.

I was later entrusted with design courses conducted with the help of Matteo Ieva, with whom I still have a strong friendship and creative scientific solidarity.

The idea that reading should be the basis of project teaching remained very clear. Students would arrive at the project through a path that started from the base building

Fig. 7.2 Graduate Laboratory, academic year 2005 / 2006. Politecnico di Bari. Redesign of the Legislative Assembly Building in Plaza Tomè de Souza, Salvador de Bahia, as an urban node. Ground floor plan and longitudinal section

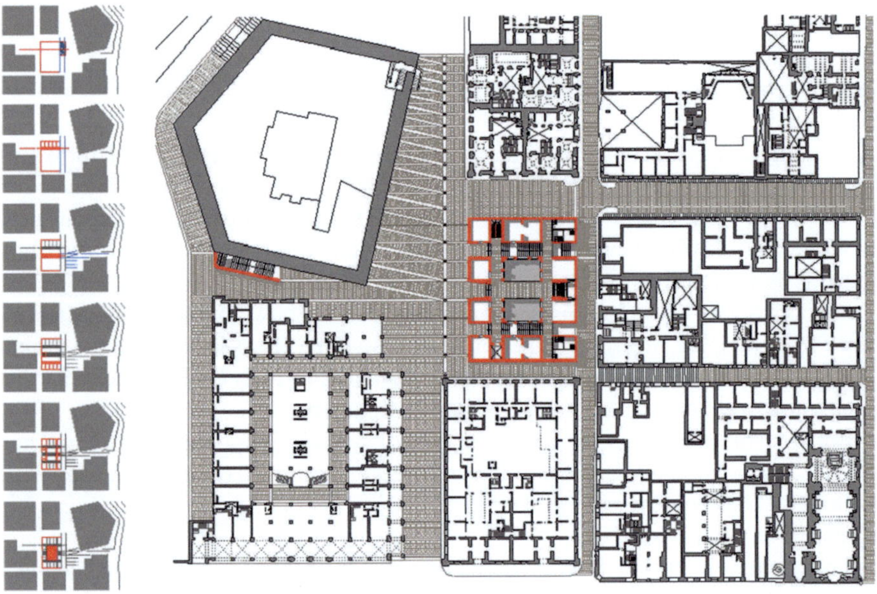

Fig. 7.3 Graduate Laboratory, academic year 2004/2005, Politecnico di Bari. Typological and design study of the planned city of La Valletta. Phased redesigns (left) of the block in front of the St. John fortification destroyed in World War II

(housing) unit and arrived at the fabric and special building by aggregations. The forming principles were identified in the concrete reality of the Apulian fabrics. The theoretical premises seemed to the students an initiatory process, with terms that were new and mysterious to them.

Then they were convinced of the usefulness of using symbolic forms, which contained, in a few signs, a set of notions and meanings. In use everything became evident. And in any case, it is true that good architecture is also an initiation which,

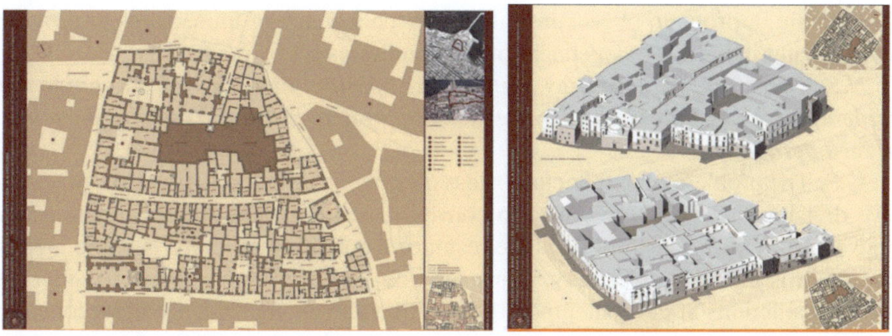

Fig. 7.4 Graduate Laboratory, academic year 2005/2006, Politecnico di Bari. Reading of the urban organism of the city of Trani

however, contains nothing esoteric except for those who refuse to understand its reasons.

Since you have asked me about the increasing sectorization of knowledge, let me quote the thought of Gustavo Giovannoni, whose figure has recently been the subject of an unexpected international interest, but also of a lot of misunderstanding.

Giovannoni argued that the loss of the center, the loss of the unity of knowledge was, even then, the result of a long-lasting crisis that, in architecture, dated at least from the rise of the Modern Movement. The new architecture, which seemed to indicate a vision of a new world, he argued, was not, in fact, based on a concordant set of theories and forms expressed through an international language, but testified, on the contrary, a fragmentation.

The existence of many discordant forms of modernity was evident, according to an interpretation that would be taken up by Saverio Muratori and Gianfranco Caniggia. The process of unitary transformation of technique and aesthetics had been interrupted, in architecture, since the end of the XIX century (and with it that principle of necessity that had constituted a fundamental ethical rule for the architect). An interruption was the obvious result of a culture of progressive specialization of labor, functional to the new social and economic balances. In *Uber die Universitats-Philosophie,* Arthur Schopenhauer had, moreover, already highlighted the functional role of university teaching to economic and political power. The continuation of this split has led to a second, current phase of crisis, symmetrical and opposite to that of the XIX century, in a condition where power is now constituted by a supranational system in which information plays the fundamental role of transmission and imposition of rules.

It is no coincidence that in many of our faculties, are proposed as examples of progressive research works by the most famous international designers that respond, with all evidence, to market demand: the very opposite of experimentation: truth constructed in vitro for the use of the *status quo* preservation. Paraphrasing a quote from Schopenhauer's writing, it could be said that today "if an idea (a project) denies the principles of the contemporary city it is considered false, or, even if true, useless."

Once again, a problem of responsibility: the inadequacy of universities to provide analysis and judgment independent of the conditions set by power arises.

N.S. Therefore, we could say that in your opinion, didactics aims to guide the student, on the one hand, to exercise the habit of demonstration and, on the other, to seek derogation.

G.S. The problem of architecture's loss of unity is a condition due to the progressive division of labor. Nowadays, to convey to students how the lost relationship between form and construction is the result of a condition of crisis, is by no means simple. It is therefore necessary, in my opinion, to do so as designers, considering the contradictions of the contemporary condition and giving evidence of its concrete usefulness. Crises are always phases of great research and renewal.

Moreover, to be concretely derived from built reality, I believe that the processual way of understanding design can free one from the weight of ideology that has contributed to the wreckage of many of the modern theories.

There are objective difficulties in teaching this demonstrative method. I believe that the school is, both, the cause and the consequence of this situation. Not just architecture schools but schools in general. The core of the problem is that students, as early as middle school, are no longer trained through demonstration. This is proved by the crisis of Euclidean geometry, which until recently formed the basis for science teaching, education through demonstration, theorems, and the use of abstraction. Gradually, in different ways, in all European countries, it was considered that this type of study moved away from the concreteness of reality and that a shift to more direct methods of observation was needed. The reality is that demonstration is no longer useful to the younger generation, composed mainly of citizens/consumers. The elite of the "producers" is formed outside of school.

This crisis of the habit of demonstration has inevitably passed into the university where students ask for practical teaching aimed directly at the profession. I would like to quote what Lucio Russo, professor of mathematics, wrote on this subject: "I shivered, listening to the argument of a student at La Sapienza University in Rome, that geometry is false since there are no real segments, as everything has a thickness. The idea was not new to me: I had already read it in the works of Sextus Empiricus; at that time scientific rationality was about to be abandoned for around fifteen centuries."

The same is true for architecture. One does not ask designers or students for proof of the project they are executing. They do not ask for the working method, rational verification, and evidence-based design. They ask for images that, in the absence of a generative process, are uncritical variants of what the market offers. The architecture student is also, after all, a consumer rather than a producer, a consumer of images. Certainly, images have their own cultural value, they are the medium of meaning. But you cannot convey that meaning if you do not have knowledge of the process that those images generated. It seems to me, then, that a good way to teach design is to use reading as a "demonstration" of design. It is, again, a circular process: you teach design through reading and you teach reading through design. Design is not only a tool for communication, it is also a tool for knowledge.

N.S. Thus the method of "redesign," retracing the forming processes as if we were designing them as architects (i.e., considering the forms not only as they were really formed, but as they could have been formed through general principles read in the multiplicity of built reality), is a tool for teaching to design in a way that is congruent with the built landscape?

As architects we know reality through design. Redesigning architecture not as historians, but as designers, basing drawings on the relationships of necessity that are established between the parts of buildings and fabrics in the city, is a powerful training tool.

I witnessed this in the feedback from my *Urban Morphology* courses at Sapienza and Roma Tre universities for foreign students who, after an initial perplexing phase, discovered an unexpected world of reflections. They realized, for example, how the coincidence of reading and design produces innovations, forms encountered in the built reality and critically interpreted, cultivated, producing some inevitable individual derogations from general principles when the architect deals with aesthetic synthesis. In other words, derogation is the consequence of a project-oriented reading, not the instrument of mannered creativity.

As you see, the eternal problem of invention comes up again in architecture, which should be posed to students starting from the actual built environment on which they will be called to work.

The texts that are supposed to explain to future architects the current modes of production are based, instead, on interpretations detached from concrete data. At their center, we occasionally encounter the physicality of architectures, the construction embedded in the great stream of ongoing transformations. Instead, we find exceptional figures of architects, their personal and heroic way of producing only seemingly new insights, "concepts" that others will translate into construction. In this framework, even the history of architecture becomes useless: it is still the consoling one Nietzsche spoke about, the "monumental history," the one of the exemplary past.

Models to imitate, after all.

It should be proposed instead, the built reality in its tangible becoming, useful for operating.

The founding fathers of the School of Rome taught the history of monuments as redesign. Vincenzo Fasolo reconstructed on the blackboard, in front of the students (with both hands), the formative process of architectural organisms, explaining how invention arose from a process. It was the teaching of history, but also of statical structures, of representation, of design.

Saveiro Muratori's "storia operante" (operating history) as well as Bruno Zevi's "operative criticism" were, looking closely, different and complementary interpretations of the common problem of reading the past as addressed to the concrete world of human actions, stratification of experiences, and experimentations on how to build and inhabited new space.

As far as I am concerned, I hope to have achieved some innovative results, in design as in teaching. To have studied in a new way, for example, some notions, such as those of *enclosure* and *roofing*, and to have introduced notions unheard for designers, such as those of ***knotting*** **and** ***substratum***, of having given a new meaning to the term "plasticity" in studies of urban morphology. Themes that I tried to convey to some doctoral students who eventually produced, in some cases, interesting publications.

As I said, I have tried to pursue my research in a secular way, taking from Muratorian teaching what is useful for designing today. Muratori was a man of integrity, but also a great moralist, tormented by the demon of ordering architecture and life with rigor. An aspiration doomed to defeat.

Built reality and its manifold irregularities do not allow themselves to be catalogued in a system of rigid rules, where everything coincides with the geometry of logical schemes and classifications. Yet the mighty Muratorian logical machine rests on still useful foundations.

That geometry of thought is necessary to understand how the city and the territory have a meaning that transcends individual phenomena, that each form is not accidental, but contains a part of universal value. It allows us, in other words, to grasp also the originality of the derogation, the richness of the many texts that constructed reality offers us, within a language (a signs structure) that generates them. Also, the fragment is an effect of which Muratori directs us to understand the origin. His eagerness for exactness, moreover, must be understood in the historical context in which he was operating, as an extreme limit to the irrational reflux that had long followed the unacknowledged ordering efforts of the pioneers of the modern.

Within contemporary relativism, this way of seeing things, seeking the origin of the built landscape even when it appears to us under the mysterious and unpredictable forms of dispersed exceptions, is an anachronism that produces original outcomes.

In my work, moreover, I have never pursued novelty. I have simply "encountered" it. I am glad when I find feedback and confirmation, which unfortunately is far from frequent, in some researcher who argues the same as I do, or who has gone before me.

N.S. I would like to dwell on two scientific activities concerning ISUFitaly and the journal U + D Urban Form and Design. Underlying these scientific activities, I recognize the intention to consolidate a link between* didactic*, urban morphology studies, and design. A combination which, in my way of thinking, would seem to be able to identify in the existing reality the reasons for the new (architectural) forms to be designed. Could you go into the merits of both experiences and, in light of recent events (conferences and issues of the journal), take stock of these scholarly activities?

G.S. These experiences are two fundamental adventures in my activity as a designer. Isufitaly constitutes the Italian section of the International Seminar on Urban Form. Today much changed, it was born from the meeting of architects from the Italian school of urban form studies, such as Giancarlo Cataldi, Gian Luigi Maffei, me, and others, and that of some British geographers who had developed the teachings of M.R.G. Conzen, a German scholar heir to the *Kulturlandschafte* studies who had emigrated to Britain during the Nazi period. With a group of his scholars, gathered since the 1970s around the Urban Morphology Research Group, we found, in the early 1990s, remarkable affinities and promising prospects for collaboration. Geography, however, is primarily a descriptive science, whereas for us the ultimate goal of urban morphology was primarily operational, project-oriented: morphology should not derive from a desire to catalogue the urban reality but from the instance of systematically understanding the built form in order to transform it.

I do not know if ours was a new science comparable, in terms of rigor, to geography, which in England had well established traditions. Certainly, it was a form of systematic knowledge, but also a critical form of inquiry, an oriented reading that proceeded through layers and stages, recognizing in the object of study the expression, even aesthetic, of a civic context. Reading was, for us, a project containing the data of reality and, at the same time, a prediction and a choice. That is why our studies could not aspire to the objectivity of the descriptive sciences. It is not by chance that there were differences of opinion when it came to finding a name for our small group of scholars. The British insisted on the term "morphology," we on that of "typology." In the end the term chosen was, indeed, "urban form." We were very lucky in this choice. The term "typology" had been so casually consumed and misunderstood in Italy that its use was, even then, a sentence of misunderstanding. Over time, studies of urban morphology became a realistic choice in an individualistic context such as that of architects, progressively aimed more at communication than construction. A sometimes-tiring choice, as shown by the ups and downs of urban morphology studies fortune in the architecture faculties (their controversial presence precisely in the Roman context, where this discipline was also born, anyway constituting for a long time an identity factor).

But it is also true, on the other hand, that identity is not inherited inertly: it is a mission in which the strenuous defense of cultural fences could be a dangerous constraint. Hence the broad meaning I give to the term "morphology." It is an inclusive term today, accounting for the multiplicity of new research investigating built form with rational and transmissible tools. The contemporary condition of those investigating urban form in transformation is one of useful disorientation. Orphans of the masters, we understand how there are no longer certainties, how it is impossible to reconstruct the lost unity of things.

The U + D journal was born in 2014. Its role, in our intentions, was to be complementary to *Urban Morphology*, an important publication dedicated mainly to urban geography, opening a debate on common themes considered from the point of view of architectural design.

The spirit that animated the initiative derived from the conviction that what the architect is concretely called upon to produce was the unfolding of an ongoing transformation of built landscapes. A choice opposed to that "art of the fragment" that was developing in Italian culture and that seemed to interpret the contemporary city condition as a set of separate phenomena, deliberately avoiding grasping the general and shared aspects that give meaning to each phenomenon.

I must say that, at least as far as I was concerned, I had ended up no longer being interested in magazines crowded with sensational images which, too much repeated, ended up making me yawn. It was, even then, a largely commercial circuit, creating myths and heroes, unexplained masterpieces and revolutions for which no proof was given, based only on a mighty media consensus.

The reason for interest in a new magazine, in a market already largely saturated, was to build an alternative to architectural publishing in which it was more important to show the immediately expendable novelty than to validate a truth, even if manifested by others.

Especially in recent years, each U + D issue is a shared project, sometime an experiment. I do not know if our magazine is better or worse than others and how much it really interests architects. I do know that it commits a lot of my time, and I think it is worth it.

I do not know how to take stock, as you ask me, of these activities. Perhaps it is too early for taking a stock, and this gives me motivation for new projects. I can only report some general impressions after so many experiences. I would say that communication is not only useful for communicating; it makes us recognize ourselves as part of the world around us: it gives shape and consciousness, in a sense, to our identity. Writing an article is not just about compiling a text. It means thinking about how it will be read, expecting feedback, comparing.

This is something I have felt the need for, ever since I was very young and used to write in Paese Sera (at the time when it was still a good newspaper) and then in Repubblica and Corriere della Sera.

The relationship with the audience was immediate. I waited with curiosity for readers' reactions, which often showed a substantial misunderstanding of the things I was saying, or concerned details that were negligible to me. For that reason, they were all the more interesting, presenting me with new points of view, allowing me to understand how others saw what I was writing. The same applies, to some extent, to the organization of seminars and conferences. It is an exchange and a comparison.

Bibliographical Note

Cataldi, G., & Rossi, M. (eds.), (2000). *Luigi Vagnetti architetto. Disegni, progetti, opere*, Alinea, Florence.

Cataldi, G., Maffei, G. L., & Vaccaro. (2002). P. Saverio Muratori and the Italian School of Planning Typology. In *Urban morphology*, Vol. 6, n. 1.

D'Amato Guerrieri, C. (2015). *Studiare l'architettura. Un vademecum e un dialogo*, Gangemi, Rome.

Fallacara, G., Restucci A. (eds), (2020). *Claudio D'Amato Guerrieri e la "Scuola barese" di architettura. A trent'anni dall'istituzione del Politecnico di Bari e della Facoltà di Architettura*, Gangemi, Roma

Menghini, A., Palmieri, V. (2009). *Saverio Muratori. Didattica della Composizione architettonica nella Facoltà di architettura di Roma 1954–1973*, Poliba, Bari.

Malfroy, S. (2011). Structure and development process of the city: the morphogenetic approach of Saverio Muratori. In T. Varena. (ed.). *Structuralism reloaded: Rule based design in architecture and urbanism*, Axel Menges, Stuttgart.

Schopenhauer, A. (1851). Ueber die Universitats-Philosophie. In *Parerga und Paralipomena*, Verlag von Hayn, Berlin.

Strappa, G. (ed.), (1985). *Attività didattica dei corsi di composizione architettonica II*, AA. 1983/84, Kappa, Rome.

Strappa, G. (ed.). (1989). *Tradizione e innovazione nell'architettura di Roma capitale. 1870–1930*, Kappa, Rome.
Strappa, G. (2023). The problem of teaching urban morphology and the ISSUM experiment. In *Urban Morphology*, Vol. 27, n. 1.

Chapter 8
Contemporary Condition

Abstract The conclusion of the conversation deals with the context in which architectural forms are produced today, characterized by an ever-increasing distance between research and the concrete reality of the built world, and by progressive specialization. Architecture is, instead, a synthetic activity. In a society in which knowledge cultivates increasingly specific fields, this could not be a weakness or an anachronism, but a great resource. A critique of the current condition is proposed: the architect should have a critical role in the society in which he is operating, but this seems not happening. The architect seems to look, rather, to adapt blindly to a condition of rapid consumption, spectacular images, and waste of resources. Architecture, from construction, has become communication. Demands for change toward a wise use of resources are, however, becoming pressing in recent years. The organicity of the built world is turning from requirements imposed by necessity in past centuries, into a conscious choice. The method of a "formative architecture" is proposed. Some personal considerations, in this regard, about the legacy of Saverio Muratori, Louis Kahn, Gianfranco Caniggia, and their relevance in the contemporary condition are reported.

Keywords Crisis · Resources · Current condition

N.S. The physical form of the built landscape seems to be a main concern in your research. It is a theme perhaps out of date in contemporary architectural production where everything seems to tend toward the virtual, the digital, and the autonomous image that finds its value especially in communication. A problem that has to do with material surfaces and solid volumes, with the perception, interpretation, and representation of the built environment. Do you think the design should be based on these material elements?

G.S. In a period of progressive abstraction in the relationship between reality and representation, investigating the concrete measure of things, considering the direct relationship between the eye that observes and the hand that draws, is a basic tool of education.

R. Salamouni et al., *Architecture in the Making*,
https://doi.org/10.1007/978-3-031-86833-7_8

The pedagogy of architectural dimensions has an ethical significance too: it is an exercise in accepting the limit that allows us to subordinate one's individual view of things to the more general context around us, to be open to dialogue with the transformations that determine the inhabited form, the anthropic space.

For some time, we have been progressively detaching ourselves from the physical relation between built form dimension and the human body. It constituted, in the past, man's way of relating to the world he inhabits. In fact, architecture was "personified," containing relations with the human body such as measuring lengths in feet, palms, and thumbs.

Measurements also tied dimensions to time, motion, and to the life flowing in the land outside the city. It was not simply the measurement of the body, but of the body in motion. The league was the measure of a distance, but in the mind of the traveler it was associated with the duration of the route to be traveled, and it indicated an hour's journey. I think this allowed the wanderers who walked the roads of the ancient world to have a material understanding of places, to feel less lost. One knew places by walking, by observing things, by touching them, and by crossing spaces.

This habit of the body with the architecture of buildings, cities, and the territories had great pedagogical value. It allowed the immediate knowledge of forms and their use. Also, it permitted to recognize the constancy of some of them, the permanence of typical and repeated characters in the built environment that man saw and named by operating, in the whole of the things he perceived, a distinction. An operation similar to what Descartes called the "distinction of reason," which separates the thing from its attributes and the attributes from each other.

But our conversation is getting dangerously close to the arduous notion of type, which I would like to avoid because of the many misunderstandings, trivializations, and prejudices that revolve around it.

I would say, in conclusion, that in the context of the crisis conditions in which our work is struggling, my research on the method of redesign is an attempt to propose didactics that link experimentation to the notion of process. A "formative architecture," in fact.

Evidently, the historic city has an important role in this. It is an open book, a text that holds the origin of language, where it is possible to recognize the rules of everyday language that could allow, even today, a contemporary design that is not in opposition to the inherited building culture. Similarly, the usefulness of studying Latin is not in knowing the rules of a dead language, but in identifying a structure that allows one to speak Italian with property, even to express and communicate innovations and ruptures. I have referred to the Roman experiences of the last century, but perhaps the practice of redesign has an even more remote and deep-rooted origin.

It derives from the familiarity, not only of architects, with historical fabrics generated from an ancient substratum and the monuments of the past, metabolized by the transformation of the city, still present and vital.

The interpretation and reconstruction of original forms has long been the main tool in Rome for understanding and producing architecture. Piranesi recomposed ancient fragments by reassembling them into a modern city. Fabio Calvo reconstructed the history of Rome not for what it had been, but for what it could have been, producing an ideal city.

Often misunderstood today, this way of looking at things has also, I believe, propitiated the formation of a new notion of "organism" as a set of rules. Historically identified, they determine a relationship between elements, structures, and systems, all of which contribute to the formation of the built environment.

Of course, this is today considered an outdated view of the world and its transformation. But I believe that the outdatedness of our work, the displacement from established modes of production in other disciplines, could be our strength. The teaching methods that form future generations of architects should take on, in fact, a reflection on the status of our discipline, rediscovering how its outdatedness opens to the future and reconsider within this framework the irruption of the many specialisms with which it tends to identify itself.

Ours is a primarily a synthetic activity. In a society in which knowledge cultivates increasingly specific fields, this is a not limitation, an anachronism, but a great resource. The work of the architect remains perhaps the only one, in a condition in which knowledges no longer communicates with each other, of which the author is (should be) called upon to understand synthetically all its aspects. Today we are rediscovering, in other fields, how valuable these lost skills are. Yet today, among architects, there is a tendency to give up this resource, to interpret even our role as a small, specialized part of a more general production system. Of the whole process the architect, in the not many cases in which he is really involved, is interested in the initial part, that of the concept. This is a contradiction that suits the large firms embedded in the communication market, but from which most architects suffer, with the progressive marginalization and reduction of their profession to a small portion of the areas historically under their competence.

This condition distances us from the concreteness of real life and progressively relegates us to a separate and abstract world. By accelerating this phenomenon, the pandemic has made it even more evident. I believe we should take this into account in a nonpassive way in our teaching.

N.S. Another aspect that connotes your interest in understanding and explaining modern and contemporary architecture is the search for a principle of belonging to a cultural area. I have the feeling that this position attempts, in some way, to free itself from the beliefs of a certain school of typology derived from the shadows posed by the "second," quite doctrinaire, Saveiro Muratori. It would be interesting to go back to the origins of your way of conceiving the study of modern architecture, leading up to your research on the architecture of Louis Kahn.

G.S. The relationship with the Muratorian School has been fundamental for me, but also very painful. The lessons of Muratori's masters, the basic notions around which their teaching revolved, were not for me a direct inheritance transmitted through courses, lectures, and research, but an arduous conquest, merging with other contributions received during my education.

Perhaps this is why the employment of many inherited principles has always occurred, for me, in a "secular" way.

I was not a student of the Saveiro Muratori courses. When I started attending the Faculty of Rome, one did not enter the classroom where he taught because the course was considered attended, for the most part, by extreme right-wing students. I later regretted that self-censorship and the missed opportunity to meet him. However, I was fascinated, and still am, by his human aspect, his epic figure of an isolated, sharp, and cursed intellectual, whom I later got to know through the long acquaintance with some of his outstanding students such as Giancarlo Cataldi and Paolo Vaccaro. The *damnatio memoriae* to which he was condemned by colleagues has always been for me a symptom of a difference that made him, for that reason alone, important.

I was not even a direct student of Gianfranco Caniggia, with whom I had instead a strong friendship during his teaching in Rome, and with whom there was a collaborative project, interrupted by his death. In the faculty, Caniggia was highly esteemed, but also culturally isolated. A spontaneous relationship of intellectual solidarity I established with him, which began from the very first meetings. I visited him for the first time on my return from a trip to Paris during which I had met Maurice Culot, who had given me a copy of *Toulouse. Les delices de l'Imitation.* Of course, he was already familiar with this book. It seemed to me that Culot's work could interest him as it concerned the relations of contemporary design to the historical heritage, and it is possible durability as a project matter. We commented on the text and discussed at length the idea of a return to the historic city as an antidote to functionalist design. As I expected, the topic engaged him, but he was completely opposed to the postmodern idea that one could build a city analogous to that of the past, borrowing the forms of history by transforming them into repertoire.

I soon realized how his reading of the conditions of crisis in the design tools had deep roots, connected to a current of thought that arose from the radical critique to the vulgarized legacy of the pioneers of the modern.

The calm passion with which he argued his arguments made me feel immediately close to him. I understood immediately that Caniggia was a genuinely contemporary architect, but of a different and countercurrent contemporaneity. What he was looking for were the laws of becoming, the rules on which to base a new, vital science of architecture, not catalogues and consolatory imitations.

At that time, he was preparing the course on special building in which he asked me to collaborate. It was a subject he had not written much about (on which he would not, moreover, publish any synthesis work) and on which he was still working. The difficulty of the topic stemmed from the fact that, while in base building spontaneous consciousness prevails (it is "building," precisely), special construction is the place *par excellence* of aesthetic synthesis; it is architecture in the fullest sense. During our many meetings, I was lucky enough to witness his reasoning on this complex topic and to follow the explanations he was clarifying for me with small sketches.

He was very interested in a publication I was preparing on modern Roman architecture for which he wrote a beautiful **and generous presentation**. He also asked me to participate in the competition for the expansion of the Faculty of Architecture in Valle Giulia. What remained of that work is an unfinished and strange project, quite anomalous in Caniggia's production. An open and problematic legacy that he left us as an invitation not to rest on certainties, to renew oneself continuously. The reconstruction that Matteo Ieva made of it, published recently, clarifies well the enigma it poses.

Certainly, that of the cultural area, about which you ask, was a central theme in Saverio Muratori's teaching. It is a reading tool that allows for "nonaligned" interpretations. Considering the unstable complexity of the built world as a product of a context linked to a specific culture, explains, for example, how modernity, presented to us by official historiography as a common neo-language, in fact is (see Giovannoni's statements) not very international. It belongs rather to the Gothic world from which derive the typical features of the Modern Movement, the load-bearing and not-closing space systems that generate the free plan and free facade. There is, instead, another pole of the dyad, equally testimony to modernity, which historiography is only now beginning to consider:

the plastic and masonry world based on a relationship of organicity between the parts. Here, construction, spaces, and architectural expression are bound together through synthetic constructive acts. That of a possible plastic contemporary architecture, is an important line of research, which I have been involved in for some time as a designer. As for the antecedent of Louis Kahn, it has particularly interested me because of the work of "recomposition" of the built world that it proposes. A great epic and antimodern theme often simplified and trivialized by historiography that should be understood in its contradictory complexity.

I am still attracted by the originality of his forms obtained by going back to the origin of the processes, the masonry power of his aesthetic syntheses due to an unprecedent way of reading history, the ancient solidarity that binds spaces and construction by generating long shadows and precipitating spaces.

These are monumental works that have lost the classical measure. They contain, instead, the fertile seed of the indefinite: the exactitude of ancient cosmogonies that has yielded to the uncertainties of the modern world. I am still interested, above all, in his idea of an artificial organicity that has nothing to do with the themes of the bizarre organic architecture crossing the modern history and beyond. Rather, in the Kahnian sense, organicity is the recognition of the characters of necessity and congruity in architectural design, the product of a unifying thought that takes place not through the study of nature or even of individual architectures, but of some universal formative structures hidden in history, in all histories.

I am attracted to his dramatic vision of the built world. It is far away from current thinking and therefore is useful. "Dramatic," I said, because Kahn was immersed in the enduring world of the American culture of industrialized seriality and liberalism, within which he dreamed of a demonstrative universe where things stand in place according to timeless (timeless, not ancient) rules and everything comes together. In a world studded with the ruins of modernity, he teaches us how we can still put them back together, give them new and human meaning. He reminisces, in this, the boy in Tarkovsky's poetic *Andrej Rublëv* who, after the many disasters and losses caused by barbarian, has not forgotten the ancient art of casting bells and is passing it on to future generations.

This operation of bringing together the fragments of the contemporary condition into a new organicity is the heroic theme that Kahn carries out in his own research. This seems to me an important aspect of the issue.

I am convinced that his research is linked, also through the teaching of Philippe Cret, to a European and organic tradition that found in Italy, between the two wars, a modern and transmissible expression. Perhaps that is why even today we understand his researches as actual and why his masonry language was so close to Italian architects when, in the late 1960s, discomfort had become evident in the face to a modern legacy. This heretical aspect of his story as a modern intellectual was predestined to have an impact particularly in the Roman milieu. Certainly, his classicism was derived from the Ancient read not in the serial, light forms of Greek architecture (as in Mies's) but in those massive, deep-shadowed forms associated with the Roman way of employing the masonry wall in continuous, opaque, massive, enclosing forms. This aptitude induced him to take new paths, to investigate, against the tide, in employing concrete in an opposite way to that of the serial frame indicated by the International Style as the very essence of modern construction. Kahn retraced, through the plastic massiveness of his works, not ancient history, but an ideal original phase of the very idea of construction. The real reason for his popularity between modern Roman architects was, in my opinion, the irruption of his forms into a context contexts in which experiments on a possible updating of the organism notion was to become. In the postwar years in Rome, the deep organic substratum of the new generations of architects, was propitiating an original critique of the international modern, carried out most intransigently.

Perhaps, this is why Kahn probably had, and continues to have, such an important authority (direct or indirect) on some of the most interesting Roman architects, such as Franco Pierluisi, Alessandro Anselmi, Claudio D'Amato, Franco Purini, Francesco Cellini, Luigi Franciosini, and others.

N.S. Your idea on architecture seems to acknowledge its ability to continuously renew itself in the sign of experimentation including, at the same time, the notions of identity and organicity.

I find this kind of interest highly topical, especially for educational purposes and, in light of the spread of current trends of teaching, directed toward breaking free from the "binding. Strength of identity"—to put it in the manner of Khoolhaas—captivated by the appeal of the inorganic, the episodic occasional and, in a broader sense, by the search for fictitious places useful to exhibit your own architecture.

The field of modern architecture can be considered the fertile ground of an experimentation in techniques of learning architectural form in its structural components and thus in its expressive character. I believe this theme of identity may also be a good topic to conclude our conversation: how this issue, quite relevant in a period of great migration and encounter between cultures, is part of the work you are doing?

G.S. As for Muratori, the theme of identity is also central in my work, but in a different way. In fact, the topic is full of contradictions, particularly in a period of great social transitions, such as the present. Ours is a very different time from the one in which the problem had been posed by the Muratori starting from urban form studies. If not defined in rational terms, the notion of identity runs the risk of indicating the harried defense of origins, a cage that hinders openness to the new and different. Identity is at the heart of exchange and dialogue, but it also fuels divisions and conflicts. Certainly, the specificities that belong to us and that we recognize identify our place in the world. However, identity is necessarily also a boundary. One either stands inside or outside. The substance of architecture is also made of continuous crises originating from what lies in between, from exchanges involving different backgrounds and customs.

It is made of hospitality, too. The architecture of everyday living has, in this, a mighty task. On the one hand, the form of the house, although in different versions, is the only common element of the Mediterranean world, in other ways composed of a myriad of identities in violent conflict with each other, expressed by the form of the church, the mosque, and the synagogue.

The house of Abraham, on the other hand, the symbol of hospitality, is present in all monotheistic religions, in the Bible, the Koran, and the Torah.

Understanding the permanence and transformation of the notion of cultural area is therefore important, but we should carefully consider the corollary to it, the presence of crisis conditions derived from areal exchanges. I think we should always think of specificities as dyads of opposing and complementary terms, admitting the infinite outcomes originated by mergers, mixing, as well as differences and resistances.

Furthermore, today we must take into account that cultural areas do not coincide with geographical areas. The distinction, for example, between the serial and the organic universe, between the plastic masonry world and the elastic wooden world is not the drawing of boundaries, but a way of recognizing the critical character of exchanges, and how they are the fertile condition that produces innovation.

The Tower of Babel is not a divine punishment (which, moreover, Genesis does not mention) for man's pride. Instead, it is a gift, the prolific chaos from which new distinctions originate, the beginning of the crisis that allows the confusion of meanings and the start of new transformations.

R.S. In this dramatic picture of the contemporary condition of design and the role of the architect, what do you consider might be a possible way forward? If architecture is necessarily an expression of the society that produces it, are we doomed to have projects that waste resources for the sake of the media spectacle?

G.S. The architect should also have, as I said, a critical role in the society in which he or she operates, but this is not happening.

On the other hand, demands for change toward the wise use of resources, toward their organic coordination with human life, are becoming more and more urgent in recent years, especially from the younger generation. The organicity of the built world, the search for congruence between its parts, are turning from requirements imposed by necessity in past centuries into a conscious choice. Perhaps, they are even becoming a new value in architecture, a discipline eternally lagging the times….

In the general confusion of things, we are beginning to understand the fundamental role played by the historically unprecedented boundless squandering of wealth in some societies of the Western world due to an availability of resources thought, for some time, to be inexhaustible.

From this organically optimistic view of the world, I try to derive my architecture.

It is possible that, in the face of new environmental crises due to wastefulness, past the naive euphoria and amazement at the changes taking place, the architecture of the future city will be composed of large empty compartments, where sometimes the footsteps of a single solitary visitor will resound, and at other times the din of the great flows of an unstable, noisy, migratory population will burst in. But the architect will not conform the spaces to the silence of one or the sonorous excess of the others. Anything that is destined to adapt to a mutation can only disregard the forms of mutation, having to possess a higher degree of solidity, capable of accommodating the different and the unknown.

The shipwreck of the myths of the "human dimension," of the many machines for inhabiting, caring, gathering, has shown, beyond any doubt, the need to separate the mutable, the accessory, and the transient (the mobile and fragile part of the built) from the solid durability of the plastic structures that support and separate. This architecture will give its exemplary products precisely in large infrastructures in contact with the crowd, the speed, and the density. In structures wherever-changing technologies do not allow form to follow unstable, ever-changing functions, but also in large nodes of territorial exchange. In all those places, in other words, where complexity comes from a lack of recognition of the connections between things. Recognizing the airport or the subway station as nodes in a structure of routes is indeed difficult in the face of the lack of physical visibility of the network itself. And yet it would be simplistic to read the airport as a fragment, the subways stations as scattered tiles in the flow of urban transformations. On the contrary, architecture must also take on the non-visible, the not directly legible organicity that binds spatial nodes together. Perhaps artificial intelligence will help us recognize the need for these new characters and the notions of organism, and congruence will reemerge precisely in the contexts from which, as recognized expressions of modernity, they would like to be furthest removed.

Bibliographical Note

Berman, M. (1982). *All that is solid melts into air. The experience of modernity*, Penguin Books, New York

Ieva, M, Scardigno, N. (2021). *L'infuturarsi della città storica. Conservazione, aggiornamento, rigenerazione, riprogettazione,* Franco Angeli, Milan

Kahn, L. (1962). Form and Design. In Scully, V. *Louis I. Kahn*, Braziller, New York

Strappa, G. (2019). The common thread that ties things together. In E. Barizza, M. Falsetti (Eds.), *Rome and the Legacy of Louis Kahn*. Routledge, London.

Strappa, G. (2021). Kahn revisited. In E. Barizza (ed.), *Louis Kahn and Venice. Tangible forms*. Routledge, London.

Strappa, G. (2021). Inhabiting the Italian historical city after Covid. In *Urban Morphology*, Vol. 25, no. 2.

Strappa, G. (2022). Cultural areas and the crisis of globalization. In U+D, *Urbanform and Design*, Vol. IX, 17/18.

Giuseppe Strappa: Poetics of the Epoché and the Formativity of Architecture

Afterword by Matteo Ieva

Writing in a few lines my thoughts on Giuseppe Strappa's rich scientific and design research activity summarized in this book is not easy for many reasons.

The first has to do with the broad themes and interests, evidenced in the multiple research paths on which he has spent his long career as a teacher and architect. These make him a sophisticated and comprehensive scholar, interested in taking a position of critical equidistance from the complex issues he deals with. As shown by the interview preceding these notes, some answers go even beyond the many stimulating questions opened by Nicola Scardigno and Rita Salamouni.

The other, more personal and therefore riskier, having for many years collaborated in his Bari courses, implies the control of the emotional component. These courses represented the opening of a new world, thanks to the intense and diversified educational experience that allowed me to intercept original and unprecedented horizons of study with respect to the background of scientific interests and knowledge acquired also through the profession.

From the first one, I will try to recall only a few points—which I will articulate in a discursive form—not being able to deepen a thought whose complexity would require extensive treatment, beyond this publication's objectives.

And yet, for the sake of temporal order, I will start with the second one, which will allow me to draw a synthetic profile of Strappa's dynamism and seriousness, as a professor at the Polytechnic School of Bari.

Called by Claudio D'Amato Guerrieri, founder of the "School of Architecture" in Bari, Strappa founded in 1993 the teaching of *Typological and Morphological Characters of Architecture,* and in 1995, he published for Dedalo *Unity of the Architectural Organism. Notes on the Formation and Transformation of Building Characters*, a monograph summarizing the lectures held in just two years of a course set with singular commitment and rigor, pioneering the origin of a new discipline on the study of architectural characters.

R. Salamouni et al., *Architecture in the Making*,
https://doi.org/10.1007/978-3-031-86833-7

The text was intense and rich in topics ranging from the theory of architecture to the study of characters and anthropic systems, investigating different scales, starting from the territory up to the built environment. It becomes the main reference of a teaching that was highly successful in the generations of students enrolled until 2012, the year of its deactivation from the educational system in Bari.[1]

A handbook for the study of characters that many have identified as an authentic treatise on architecture, as D'Amato was often fond of saying, capable of transcending the didactic purpose alone. It demonstrated the interest in fusing together documentary research and operative criticism, in which the scholar's aptitude for rereading emerges. It proposed a new and original interpretative key, the theoretical achievements of the main figures who have dealt with the characters of buildings[2] and, at the same time, the deep layers of research on the components that explain the structural manifestations of man-made space, without neglecting the critical contribution of the major exponents of modernity.[3]

The success of the teaching, I believe, can be attributed first and foremost, to the clarity of the content that gave students the opportunity to acquire useful knowledge for interpreting the world through principles based directly on reality. At the same time, it allowed them to enable the acquisition of a solid foundation for recognizing and designing the multilayered and complex housing and special types, decoding urban phenomenology through the recognition of its own structural laws (of formation and transformation) and hierarchies planned and/or generated over time, inducing an exegetical effort to understand the changing meanings on which our field of interest turns in contemporary times.

A task that was based on the powerful background of interests cultivated in his long career as a researcher, inspired no doubt by the experiences nourished by a solid training acquired at the School of Rome in the second half of the 1960s. Also, he did not lack the relevant contribution of studies on urban form and building typologies carried out, among others, by such decisive figures as SaverioMuratori and Gianfranco Caniggia. Strappa will base much of his speculations in his mature years on the legacy of experiments by the two Roman masters. He demonstrates an undoubted closeness to their school while sharing the "scientific" foundation of the hypothesis of method applied to the reading of reality. He expresses a thought constantly constructed on the plane of dialectical opposition of phenomena.

See, in his work, the fruitful research on the dyad of opposing and complementary terms and, with the questioning of many dominant currents of thought, the investigation of the principles that aim at the concreteness of the built environment. He

[1] For more on the temporality of teaching and alternating teachers, see Nicola Scardigno's interesting article (2021, 156–161).

[2] Noteworthy among them is the original openness to the little-known figure of Enrico Calandra, a professor at the Giovannoni's School of Architecture in Rome with whom the first generations of Roman architects were formed.

[3] The density of the themes presents and the relevant documentary background that innervates the book's structure have also been frequently referred to by foreign scholars, who have pointed out the desirability of publishing the English-language version to enable dissemination in the international context.

renounced what is abstract or ideal, refuting, in opposition to a widespread tendency, any form of hypostatization of ideas about architecture since he does not platonically attribute to them an ontological consistency.

For the purpose of arriving at an impartial reconstruction of his theoretical courses, it is not irrelevant that his adherence to the goal of the so-called Muratorian School, far from dogmatic, stands with a productive eccentricity. It surpassed the legacy that had connoted the areas of interest of the School's two main exponents, who had focused their observations mainly in the field of pre-modern architecture.[4]

In fact, Strappa studies with passion and scientific interest the dynamics of the International Modern. He researched "secularly" without idealizing (one might say) frantically its importance and historical scope and grasps its essential features by reading his course not as a rupture of the past codes but as a continuity and above all, a renewal in design terms. Many are his publications on the subject that thoroughly analyze the characters of the two distinct geographic-cultural areas. They range over the lineaments of so-called Mediterranean modernity, in parallel with the expressions that had generated the architecture of the other side of the Alps, manifested with original innovation and distinct coherence.

Moreover, the attention to the values of modernity has also found plenty of space in the educational activities transmitted with the courses of *Architectural and Urban Design* delivered in the long years of teaching in the Faculties of Bari and Rome, nourishing the knowledge and stimulating the curiosity of the students.

The characteristic and interpretative trait present in Strappa's lectures on some typical organisms of the International Modern were always analytical. They exemplified an authorial complexity differentiated by cultural spheres and tending to search for the structural and archetypal laws of a "linguistic" truth.

It resonates with extraordinary evidence in the atlases, monographs, and published articles,[5] but also in the extensive citations to his writings that are added to the constant manifestations of participation of the international scholarly community in his critical thinking.

---

[4] It is well-known the Muratori's refusal in still feeling modern: "I have had to make considerable effort to shake off the commonplaces acquired as a child of youthful modern wishful thinking; I have employed all the experience from my twenties to forties to identify the unresolved problems of present culture; from forty years onward, with the study of the urban fabric of Venice and Rome, I came to understand the laws of the typicality of urban forms and the cyclical nature of the world of the city, as well as that of man, I employed another ten years of work on the question of territory, and finally I tackled the problem of self-consciousness, that is, the adventure of civilization."

Caniggia, while initially focused on the themes of modernity, later discovers in the formative processes of anthropic organisms a creative path that will nourish the scientific aspirations practiced through teaching, leaving out the initial tendency to grasp in that historical phase the prodromes of a real post-modernity: "It is thus SaverioMuratori who authentically reveals himself as 'post-modern,' not in the stylistic-morphological sense, but in the more general aspect of a critical rethinking of the subversive unreality of the modern era and yet solidly taking his start from the rational and systematic, the logical control of architecture that is the afflatus and authentic legacy of the Modern Movement."

[5] Of particular note, along with those published in collected works, conference proceedings or for the newspapers Corriere della Sera and Repubblica, are the editorials written for the journal U + D Urbanform and Design of which Strappa is editor.

After all, this is a cognitive perspective, updated and revolutionizing, of an undoubtedly original work, in which reflective and scholarly trajectories originated from a strong theoretical structure.

A quick mention to the most significant ones, recalled by means of a few key words, will help clarify the other point proposed in the opening.

cultural areas, operant history, process, interpretation, reading, form-structure, subjectivity, form, type, formativity, project, utopia

A persistent line of research that was completed in the progress of time with the studies carried out on some European cities. It deepened the cultural diversities theme, mirroring the different architectural-linguistic civilizations, analyzed with the observation/reading of the characters of the built environment, selectively researched, in concomitance with the identification of the essential differences and the laws of determination of these distinctions. This codifies in the recognizance of the two main cultural areas: the Mittel/Northern European one with a *wooden-elastic* character and the Mediterranean one with a *masonry-plastic* character.

Specific qualities were brought as a reasoning on the contemporary without any peremptoriness, as they are read in the light of the structural variations they reflect in today's condition.

It is not by accident that the perspective toward which he moves shows an indubitable, integral impartiality of opinion toward the expressions of architecture of all times. He was not surely seeking in these the image reflected in a sphere of judgment conditioned by dominant thought or interpretative myths that have traversed the historical course with the aim of restoring a possible meaning—often translated into the "restoration" of the unspeakable aimed at filling the gap. Strappa, on the contrary, thoroughly investigates the essence proper to the phenomenon that generated the "fact" retracing the entire process that leads back to the origin of what is manifested.

Therefore, he pursues a critic-interpretive dimension based on the assumption of the "storiaoperante" (operant history), already practiced by Muratori, in which the individuality of happenings is explained through the relationship that comes to penetrate the interdependence between: expression of the phenomenon, historical course, and spatiotemporal conditions.

An interpretation promising, from the perspective of the Foucauldian *historical* a priori by which one "searches for the shadow that the questioning of the present casts on the past," a momentum toward a possible advancement and a hypothesis of a future that looks to the project.

In this way, it expresses the ability to place itself dynamically in the process, interpreted as a synthesis of moments defined by a double polarity, that of data transmission and the other of the interpreter with his vision framed in history.

This scenario of critical reflection on the problem of interpretation seems to propose the idea of the conjunction of the different moments through the so-called history of effects, which is nothing but the *Wirkungsgeschichte* proposed by Hans Georg Gadamer, according to which all past interpretations mediate the interpreter's pre-understanding of the interpreter without being aware of it, allowing to reach the so-called fusion of horizons.

Many of the issues proposed in this reasoning, as can be seen by scrolling through the interview, hinge precisely on the theme of interpretation, which constantly appears in the large production of essays and pre-readings intended to design.

On this point, notations in agreement with Luigi Pareyson's thought recur frequently. For example, a constant present in the assessments of the works, the outcome of art making, refers to the idea that also identifies the productive and contemplative aspects.

This is a recognition of an operative "forming form" (idea to interpret) with a view to questioning the hermeneutical basis of existence, in relation to the possibility of arriving at an inseparable linking of what descends from spontaneity with the ability to transpose, the system of the relationship of hermeneutic circularity, precisely, the system of the relationship with the conscious self together with that outside one's consciousness. Mode of exercising an exegetical consciousness in a non-self-referential form constitutes the logical order leading to city studies. The "reading" of the existing environment of its visible and temporally and spatially variable form structure, revealed through its own laws, supports a verified method of inquiry that it is never framed in a predefined classificatory framework and opens to the almost indefinite interpretive possibilities.

As revealed by his ability to provide himself with "tools" that vary continuously because of a "truth" that the being (territorial, urban, or built) announces ontically from time to time. And this is demonstrated by the vast production of studies on ancient and historic cities, such as Rome, Trani, Castel Madama, Castel Di Sangro, Bitonto, Carezzano, Partinico, Ponza and Ventotene, Montesilvano, Mesola, Urfa, Porto, Salvador de Bahia as well as interesting modern and contemporary cases, such as the eastern suburbs of Rome, the northern suburbs of Milan, Latina, Sabaudia, the large commercial service areas in Quebec City, the Diagonal in Barcelona, Tlaquepaque in the Guadalajara region in Mexico, and the brownfields in Mendoza, Argentina.

It is, after all, a method of inquiry into reality that Strappa employs with regular systematicity in every field of observation by resorting to an authentic substratum of an epistemological nature. It is integrally supported by a self-conscious thought that subordinates the subjective vision permeated by an inverted doing on the fullness of the *ioassoluto* (absolute I), tending to express a creative conception comforted by the universal and exclusive ego, capable of inspiring its own action of modification only in relation to its own thought. A view that, of course, does not lead to a complete denial of subjectivity because it is aware that existence is in the individual and not in universal concepts. A man, by virtue of this attribute, cannot prescind from his own subjectivity. Consequently, the thought, if considered in its abstractness and universality, cannot understand the singularity of the reality.

Among the many writings that illustrate the scope of Strappa's speculative practice in which the concepts recalled are substantiated, it is necessary to remember the monograph "*Architecture as Process. The masonry-plastic world in the making*" (Strappa, 2014). A text of maturity, this work addresses different lines of research demonstrating the interest to follow the post-Husserlian phenomenological postulate of observation of the world in the search for the dual relationship between the *I-demand (question self?) and the object and the I-demand and the subject.* Aware that every act of consciousness is not pure subject—the one who looks at reality—but is an intentional attitude, in the sense that every individual observation, especially in the field of architecture, in the most recent trends, is a continuous relation to one's own consciousness. But it is precisely due to the supposed dualism that generates division between seeing the world as sensible experience and the consciousness that constructs the meaning of the thing that is vigorously opposed, placing on a synchronic data, therefore, the material and the formal content.

This aspect of his thinking the way in which man relates to the concrete world is clarified in the reasoning built around the **second question** posed by Scardigno. And still further on, reflecting on the foundations of experience and the concept of form, he states that matter cannot be given without form because it is already form and this is within matter.

The extraordinary lecture on the *matter-material-element-organism*[6] sequence is the clearest demonstration of this because it offers an original interpretation regarding the impossibility of arriving at the separation of the sensible and the rational. It makes use of the mediation of man and his intentionality for the end to which the act itself aspires in the production of the thing.

This component of judgment determines the hypothesis of a kind of "ontology of pure forms ",[7] considering that one starts from the forms or multiple ideas that constitute the ideality of the thing and arrives, inextricably, at the understanding of the essences that are already proper to any phenomenon. Also, based on one's experiences, at any time, it can construct categories on an interpretive basis. But the critical readings proposed by Strappa, hinging on the interrogative question of how to overcome the problem of the intentionality of the subject judging phenomena, suggest that the answer lies in the ideal turn of the primary reconstruction of the

---

[6] Regarding this aspect of the research, one could quote a suggestive reasoning of Heidegger on the "thing" that summarizes, with the well-known method practiced by the philosopher, some passages from Strappa's paper "The Sequence of Tectonic Transformations," taken from (Strappa, 1995, pp. 53–75):

> "One thing, for example, is the pitcher. It is a vessel, a thing that contains in itself something else" (1976, p. 110). "What is it that constitutes the thing as a thing? [...] What is proper to the essence of the jug is never manufactured by production. [...] When we fill the pitcher, in the filling the liquid flows into the empty pitcher. [...] If, however, the containing resides in the emptiness of the pitcher, then the potter, who forms the walls and bottom of the pitcher on his wheel, does not properly manufacture the pitcher. He only gives form to the clay. Indeed no: he gives form to the hollow. But is the jug really hollow? Physics assures us that the jug is full of air" (Heidegger 1976, p. 112). Emptiness for Heidegger (ed.) "contains in two senses: taking and holding - and creating unity, connection between them - [...] unity is given by the act of pouring" (p. 114), to pour is to offer, "the being-brochure of the jug unfolds in the offering of the poured."

[7] Scalarly, these can refer to the urban area and the aggregate system connected to the building system of base and special types.

genesis of form and its process through recourse to *epochè*, which allows one to be free from any pre-constituted belief. In other words, liberate one from prejudgment or pre-categorical implication, as overcoming the method based on empirical experience alone.

Describing the phenomenon as it is given and inside the limits within which it is given to the subject, one might say, transcendental in that one seeks to grasp the thing from its origin by means of a pure gaze: interrogating the phenomenon and grasping it in a continuous nexus of moments that describe its genealogy and historical-processual bearing.

Hence the wide range of studies he undertakes in the different phases of the unfolding of a thought that constantly places at the center of research the dialectic of opposites—as we said—projected toward a "place for the possible" capable of unveiling the phenomena of antithesis and conflicts in being in the world of building and urban structures. The goal of knowledge of operating read with the point of view of the becoming of the forms produced over time by man and diversified in the distinct geographic-spatial environments. The originality of the method employed consists, not surprisingly, in the ability to perform a synthesis based on the interpretation of "form as organism." A concept related to that of type and defined already in the 1995 volume, on which he builds the line of research practiced also in the field of urban morphology.[8]

Moreover, on the concept of *formativity*, related to that of *generativity* recalled in this paper, Strappa devotes a series of interesting considerations noting that it contains within itself both the technical and the artistic aspects, and, while acknowledging the differences between visual arts and architecture, he dwells on the theme of *form* noting how it never originates from nothing, but is the result of the rich accumulation of experiences that precede the project, transformed into the built world through the device of consciousness.

A judgment that overcomes Benedetto Croce's idea in which invention precedes execution and form itself anticipates the activity of the operator. At the same time, fully sharing the position of the Turin philosopher, he judges (in the formative process) the effectiveness of the critical-intellectual ingredient as the soul of the component of spirituality, no longer understood as pure feeling. On that ingredient, he delineates the individual coefficient transmitted with the project that gives the architect the possibility of modifying/transforming reality responding to the needs and the continuous civilizational changes.

He writes in this regard

> *"…How does the architect contribute to this process of transformations? Certainly, through the design, which should, however, be understood as a renewal of what already exists. The project prefigures the provisional completion of a process and should be integrated as the concluding, future-oriented phase of ongoing modifications. It possesses certain internal rules that govern its formation: it can be understood as a kind of "project organism. "In this sense, a central role is played by the notion of formativity …"*

[8] A scholar of urban morphology, he has built an unconventional thinking on this line of research with which he has contributed to growing the notoriety of the Italian School of Morpho-typology.

On this theme, of great importance to those who practice design, the position appears clear and unequivocal. The project, while contemplating a condition of inescapable and necessitated authorship, denoted by the stylistic formulary of the practitioner, is not "imaginative creativity"[9] but "creative power"[10] and, as such, should be framed in an idea of process in the making.

The theme of the meaning of the "words" essential to the project is here extensively sketched both in Scardigno's presentation and in the articulate reflection that appears in the responses. An aspect of some complexity, on which I would try to spend some consideration instead, is captured in the challenge to the individual thought of the worldview that Strappa launches in suggestive terms: project vs. utopia.

"The architect should possess, in other words, his own vision of the built world in which to place things and make sense of them. This vision should contain not only the perception of our surroundings. It should contain the myth and utopia that detach us from the everyday contingent and allow us to investigate into new territories."

There is no doubt that the utopian component governs the idea of something that lives temporally beyond the present. But, if the term derived from the Greek οὐ and τόπος means "place that does not exist," one wonders if one can give place to that which has no place. Utopia, after all, as a form of subjective anticipation and projection-representation of a tangible and realizable future is moreover ambiguous. The anticipatory character is present, which is based precisely on something other than pure hypothesis. And this other is the project.

Marc Bloch speaks of concrete utopia being opposed to abstract utopia, as if the project could make it factual, but with the contradiction of erasing the specific postulate of utopic which is openness to novelty and, I would say, also to imagination, to invention. As Strappa intuits, the project cannot renounce to frame itself also in this sphere that is not exhausted in the prevision act since a beyond is always proposed.

In the field of architecture, Bloch's problematic judgments, plausible for others disciplines, are not matched since utopias have met so far an opening only in manifestos and programs, often launched in extreme forms, in other cases acting in interpretive form.

After all, utopia is generally proposed as something ideal, a theoretical horizon that is not fully attainable. From this point of view, it is an extraordinary instrument of thinking and imagination that allows one to "desire otherness" with respect to a spatiotemporal experience limited to the conditions in place at their own time.

And therefore, to the question of what is really utopian and expressible in a project, one can think of the need to hopefully pursue the overcoming of the material conditions on which the present is nourished, launched toward a possible otherness, trying to overcome the disenchantment of the abstract projected into the sphere of the inadmissible.

---

[9] Creativity is often understood by art and architecture critics with a negative connotation in the sense of deciphering it as an opposition to the already made.

[10] As suggested by Caniggia and Maffei (Caniggia G. and Maffei G.L., 1984), to be interpreted according to the meaning ascribed by linguists to denote the individual's ability to use the language in his own right, individually, by implementing his own *parole*.

This is the paradigm with which Strappa gives life to his projects but with a different emphasis that, in my opinion, correspond to three moments of improvement of his thought.

The first coinciding with the experiences of the fertile season of projects ranging from 1972 to 1987, in which the specific component appears more pronounced toward the search for drawn forms from history without feeling the resistant sieve of the typological perspective. It is worth mentioning only the project for the Terni Cemetery, which manifests a series of interesting references, mostly referable to the characters of the Etruscan world burials with the irreducible mediation of an active and powerful modernity.

The second is to be framed with the phase of teaching in Bari. The expectations of the School proposed by Claudio D'Amato on the organic world and the use of stone are fulfilled, by means of a continuous dialectic of exchanges of opinion. Adding to this, in parallel, a stable interaction with some scholars of the Muratorian School. The decade that ends in 2007 with the move to Rome corresponds, roughly, to the moment of theoretical and design research that produces experiences of adaptation and updating of building types in a perspective of renewal in contemporaneity. This, today, thanks to Strappa's interest in the Modern, is enriched with new ingredients.

Looking at the competitions for the Rest House in Trani or the one for the Tor Tre Teste church in which his friend Gian Luigi Maffei, an important figure in the Muratorian School, collaborated.

In the last one, I think the confrontation with the Roman School should be acknowledged, which will offer a valuable opportunity to cultivate other interests and a challenge that will take him into a field of experimentation only partly practiced before, as the competitions for Carezzano, S. Marinella, Tlaquepaque, etc., demonstrate.

But it will be above all the PRIN research on Palermo's Zen and in particular the solution for Expo after Expo, each with different degree of complexity and with a high gradient of unknowns, that will show a different degree of design "utopia" reverberating on the paths of a new, updated and convincing avant-garde.

And these paths Strappa still runs today with a deep sense of ethicality, whose practice inspired lyrics shunning the din, are, to paraphrase Heidegger "Places of stillness" … in which one glimpses "a dark light that does not deny clarity but that excess of brightness that prevents sight."

Bibliographical Note

Caniggia, G., & Maffei, G. L. (1979). *Lettura dell 'edilizia di base.* Marsilio, Venice.
Caniggia, G., & Maffei, G. L. (1984). *Il progetto nell'edilizia di base*, Marsilio, Venice.
Heidegger, M. (1976). *Saggi e discorsi*, Mursia, Milan (or. ed. 1936).
Ieva, M. (2018). *Architettura come lingua. Processo e progetto.* FrancoAngeli Editore, Milan.
Pareyson, L. (1960). *Estetica: Teoria della formatività.* Zanichelli. Bologna.
Scardigno, N. (2021). Morfologia urbana e progetto. Il rinnovamento del pensiero tipologico-progettuale e la didattica della Scuola di Bari. In *U+D Urban form and Design*, special issue: *la ricerca di morfologia urbana in Italia. Ripensare la morfologia urbana. 1 Tradizioni e nuove scuole*, n. 15.
Strappa, G. (1995). *Unità dell'organismo architettonico. Note sulla formazione e trasformazione dei caratteri degli edifici*, Dedalo, Bari.
Strappa, G. (2014). *L'architettura come processo. Il mondo plastico murario in divenire*, Franco Angeli, Milan.

R. Salamouni et al., *Architecture in the Making*,
https://doi.org/10.1007/978-3-031-86833-7

MIX
Papier aus verantwortungsvollen Quellen
Paper from responsible sources
FSC® C105338

If you have any concerns about our products,
you can contact us on
ProductSafety@springernature.com

In case Publisher is established outside the EU,
the EU authorized representative is:
Springer Nature Customer Service Center GmbH
Europaplatz 3, 69115 Heidelberg, Germany

Printed by Libri Plureos GmbH
in Hamburg, Germany